In Search of the Unknown

WILLIAM JEVNING

Copyright © 2017 William Jevning

Second edition

All rights reserved.

ISBN: 10-1522749535
ISBN-13:978-1522749530

DEDICATION

To my old friends with whom I began this journey:

John Adams

Mark Barshaw

Jim Dawson

Kirk Goddard

Brian Godwin

Scott Martin

Milo Rogers

ACKNOWLEDGMENTS

There are so many people to thank for working with me over the years, first I wish to mention the people that worked on editing this book, Chandra Bryan and special thanks to Tim Gormann who did the final edited version. I also wish to pay special tribute to the members of the Board of Directors of the former PCSIT: Donna Rae Querry, Alice Nicholson, Ron Madler, Debbie Madler, Candice Philpott, Don Turner, Carol Szimonisz, and Jack Livingston you all were very special friends and I could not have done my work without you.

I wanted to do a second edition of this book since the first printing. This is not a book about Bigfoot, but rather how I came to become involved in the subject of Bigfoot. This covers my entry into the subject of the Sasquatch and follows some of the memorable events (from my perspective) of my experiences until I began writing my first book Notes from the Field, Tracking North Americas Sasquatch and the creation of the Jevning Research Group in 2003.

William Jevning

EARLY YEARS

I was four years old when my eleven-year-old brother Buddy was killed suddenly in a tragic accident. His best friend was showing him the .25 caliber pistol his mother kept under her pillow. Somehow the gun went off, and the bullet went through my brother's heart killing him instantly.

We lived then in the small town of Orting, Washington, which is situated in the Puyallup valley between the Carbon and Puyallup rivers. This is where my journey began. Orting in the early '60s was a small town of approximately fifteen hundred residents. It was a nice, quiet community where most people knew each other.

After my brother died, an old family friend, Charley Welcome started taking me with him on fishing trips. At the time, I had three older sisters and one younger sister, plus my mother pregnant, so I am sure this was Charley's way of helping my grieving and stressed parents.

As a very young boy, I had absolutely no idea of what the world was like outside of that small town, but my world view would soon get much larger.

Charley had been raised in the small logging town of Carbonado, Washington. He was well acquainted with the forests of the Pacific Northwest. In fact, he worked for the National Park Service at Mount Rainier at the time and spent as much time in the wild forests as possible.

(William Jevning age 5)

When Charley began taking me on fishing trips, I was often too small to walk far. He would often carry me on his shoulders, especially when we needed to cross the Carbon River using railroad bridges.

As a very young child, my parents wanted me to be very aware of everything in my surroundings on these trips. There were plenty of wild animals to be wary of, not to mention the possibility that I might trip on rocks or fallen tree branches and tumble into the river.

My father and Charley had a unique way of teaching me to pay attention to my surroundings and where the adults were at any given time. They pretended sometimes that I had disappeared. This would infuriate me, so I stayed closer than ever to them.
This may seem a cruel thing to do to a small child by today's views, but I am very thankful for the lesson as it has served me very well to this day.
As these fishing trips grew in number, I began to feel very comfortable in the outdoors. Charley was always teaching me something about the rivers, trees, and plants, and he showed me all the animal signs we encountered.

Over time, Charley and my father taught me many valuable lessons about the forests and wildlife. Eventually I would take my place with them fishing and hunting, and these were things I needed to know well.

When I was eight years old, we moved to a 40-acre farm outside the town of Orting, and my whole view of life changed. A man from Tacoma named Richard Whitman owned the land and leased it to my parents. He was an elderly man whom we all affectionately referred to as "Old Dick." These 40 acres spanned the Puyallup River, having 30 acres on one side and 10 on the other. The house, barn, and outbuildings were on the 30-acre side. Old Dick's property was located behind a state nursing home called the Soldiers' Home. This had originally been built for veterans of the Civil War and housed mostly World War I veterans when we lived there in the '60s.

(Charles Welcome)

We lived on Bell Road, which had only three homes on it in those days, ours being at the very end. The road was gravel and dirt mostly, very rural. It was just the kind of place to encourage a boy's growth and curiosity.

My father asked a friend and coworker of his to bring his bulldozer to our new home and make rough roads throughout the acreage. Much of the land was inaccessible otherwise and my father wanted to be able to use it.

(My parents in the early 1960's, "John and Gerry")

Many of my fondest memories from childhood are of my first collie, Boy, who was my constant companion, and of my BB gun. I shot so many BBs my father always joked that we could mine copper there. That was a great place to live, and I am lucky to have had such a place in which to grow and learn.

My parents did well financially, but were not the type of people to spoil their children. My father said when we moved to Old Dick's property that he was not going to buy us a lot of toys. He said, "There's forty acres. Do it!" When not in school, my two younger sisters and I spent the majority of our days building and playing in the dozen or so tree houses we built.

My three older sisters had already gotten jobs and were married by the time we moved to Bell Road, so the only companions I had to adventure with were my younger sisters. Sometimes a friend from school would come and stay the weekend, but this was not often.

We could not resist exploring the roads my father's friend had built through the forest. We got scolded sometimes because my parents were afraid a bear would harm us or that we would experience some other dangerous accident.

The river was not far from our house, and there was a large pond that was mostly swampy next to the river dike. We used to go there whenever we could to catch frogs or play on the small raft that we had built.

To get to the swamp, we walked about two hundred yards along a very old moss-covered road that led through the forest to the river. Once we reached the swamp, there was a trail that led down approximately 10 or 12 feet to the edge of the water.

One particular summer day I was leading my sisters there as we planned to play on the raft and possibly go to a small island in the middle of the swamp to explore. As I reached the edge of the embankment leading down to the edge of the water, I came face to face with a large black bear. There was only about five feet between the bear and me, and all I could do was yell, "Bear!" and "run!" All three of us took off running as fast as we possibly could for the house.

(July 4th 1968 at our home at "Old Dick's property")

I thought the bear would catch us soon and had to be right on our heels, but it was nowhere in sight. I was very relieved, especially since we had been told never to go there without an adult.

I told our mother what we saw, and she said to stay inside the house until our father got home from work. When he arrived, we told him what we saw. Charley had moved onto the property with us at that time and had a trailer he lived in not far from our house. My father went over and told him a bear was on the property and that they needed to get rid of it.

I was about nine or ten then and it was all very exciting, except for the fact that I could have become lunch for the bear!

(July 4th 1968 at our home)

My father was an experienced hunter and fisherman and so were many of his friends. He called one such friend who owned hunting dogs. They planned a big hunt to get that bear. I recall the first day of the hunt, which began just a day or two after our frightful encounter.

My father wanted me to go along with the men so I would learn about bears and hunting. I went to the place we saw the bear along with my father, Charley and my father's friend who brought the tracking dogs.

Another of my father's friends kept honeybees on our property to produce honey to
sell. He had nearly 300 hives. Some of the hives were located near the place we encountered the bear and looking back, I am sure that the bear was interested in the honey and not three children.

(July 4th 1968 at our home, note a ten year old William Jevning left)

Several of the hive crates were torn apart, affirming the fact that the bear was there for honey. The men agreed that the bear must be found and shot, or it would continue to destroy the hives.
We split into two groups, my father and the dog handler in one, and Charley and me the other. My father said Charley was an expert hunter and would teach me a lot about bears and that I should pay close attention to all he would tell me.

It was all very exciting as we conducted the hunt for several days, but without any luck. The bear had apparently left the area, honey or not.

(July 4th 1968 at our home)

We found many tracks, feces, and trees clawed by the bear. Charley carefully explained all of this to me. I was highly motivated to learn all I could about bears after my close encounter with that one so I could avoid such an experience in the future.

This event taught me to keep my guard up throughout the rest of my life. Whenever I am in the forest, I always stay alert to all animal signs, especially for the presence of any potential threats.

Charley thought that after the bear hunt it was time I had a better education regarding wildlife. He was friends with a man who trapped wild animals unharmed when they wandered too close to people's homes. He would take them far into the mountains near Mt. Rainier and release them.

In Search of the Unknown

One day the two of us went to visit this gentleman, and it proved to be a lasting experience. When we drove into this man's driveway, the first thing I noticed was the many pens and cages he had. Most of them were occupied with raccoons. I had seen raccoons on television, but never a live one.

Those animals mostly seemed vicious, and I was afraid to go near the cages where they were kept. A few seemed friendly, but Charley told me not to get close as they did not know if they had distemper or not. One particularly nasty animal in a large closed-in pen was a bobcat, and to this day I hope I never encounter one close up in the wild. That animal was the meanest creature I had ever seen. It reminded me of the cartoon Tasmanian devil.

There were a few raccoons roaming around freely, and the man there said they were his pets. They seemed as friendly as house cats, but I still kept my distance, even with them.

Both Charley and his friend told me as much about these animals and other wild creatures as they could, and it made a lasting impression on me.

My father and Charley would have me carry their hunting rifles whenever the opportunity arose. They wanted me to get the feel for them, and they taught me everything about them. I grew up with a great respect for firearms.

Charley taught me to fish before he and my father felt I was old enough to go hunting. My uncle had given me a fishing rod for my tenth birthday and when fishing season began that year, Charley decided it was time I learned.
We would dig our own worms behind the barn the night before a planned trip, then go to either the Carbon or Puyallup Rivers. The rivers then were full of trout, and we always got our limits.

Charley did not believe in purchasing anything except fish hooks and sinkers. He rolled his own cigarettes, and we used the empty tobacco cans (Prince Albert) for worms that we put in our shirt pockets. We even broke alder branches in which to carry our fish.

We normally did not fish lakes, preferring instead the rivers and creeks. We would begin early in the morning and fish the first half the day upstream, then work our way back to where we started. Charley said this was so the fish would not see us, as the sun would be at the right angle to hide us. It worked every time.

Growing up near the forests, we had never heard the word "Bigfoot." Even now I am not sure if there were any such creatures in that area then, but there were two very odd incidents that happened when we lived there.

Sometime, either late during the summer of 1968 or early that fall, something odd happened one evening while we were watching television.

I recall seeing out our front window a light coming up our road. We lived some distance from the nearest neighbor who was an elderly lady named Mrs. Jackson. She lived alone, and she rarely had visitors aside from my sisters and me, so it was odd to see a single light making its way to our house.

I told my mother what I had seen and she said we would soon find out who it was. There was a loud pounding on our front door a few minutes later, and my mother answered it.

There stood a very scared looking boy about my age. My mother urged him to come inside and asked him who he was and what was he doing here.

(Sketch courtesy Jerry Bishop)

The Orting school district was very small so I would have known any boy my age. However, I had never seen him before, so thought he must be from somewhere else. He acted like he had been terribly frightened by something, and it was some time before my mother was able to get him calm enough to tell us what happened.

It was well after dark, and my mother asked him what he was doing out on his bicycle so late. At first the only thing he could say was that the "Rock Quarry Monster" had chased him, and he kept repeating this over and over. Finally my mother got him calmed down enough to tell us his name. He said he was staying on the other side of the Soldiers' Home with his grandparents for a few days.

He explained he had been out riding his bike and must have taken the wrong road and got lost. That's when he thought he saw a man standing next to the dirt road we lived on. He approached the "man" to ask how to get back to his grandparents' home. As he got close to this person, he realized it was no person. It was a huge gray or white hair-covered monster.

He said it came toward him and that he was afraid it was going to eat him so he pedaled as fast as he could away from it and ended up at our house.

He explained that he had heard of the Rock Quarry Monster but didn't believe in it until now—that is, until it almost got him. My father had been in bed reading a magazine, and came into the living room about this time. He asked who his grandparents were. The boy told him, and my father said he knew them and would take him to their house.

My father put the boy's bike in the back of his pick up and took him home. I asked my mother if she had ever heard of this Rock Quarry Monster. She said she had not, and his getting lost in the dark must have scared him into believing he saw something.

This sounded reasonable to me, and I didn't think about it until I overheard my parents talking later.

Before I went to sleep that night, I heard my mother ask my father if he knew anything about this Rock Quarry Monster. The county owned and operated a granite rock quarry not far from our home. The granite was used to line the rivers in the region to help prevent flooding. He said some of the residents around there had claimed to have seen some sort of gorilla looking animal and had dubbed it, "The Rock Quarry Monster."

My father said they were "nuts", adding that it was probably an old drunk who was a hermit that poached deer near the quarry. I thought my father knew what he was talking about and that was the end of the matter. However, I couldn't help but wonder if it was just an old hermit, why had that boy been so scared and insistent that he had seen a huge hair-covered monster? I also wondered why an old drunk hermit would be in that swampy wooded area late at night with no flashlight. The boy's description sure did not sound like a person to me.

There was another odd incident that happened during the summer of 1969 that still haunts me to this day. I was riding my bicycle on the long road that led to our home one day, much as I did most days during that summer. I would often ride over to the Soldiers' Home, and visit with the old veterans there. I listened to their stories of action in France during World War I, or other stories they enjoyed telling a young boy.

Following my visit that day, I arrived home and saw my mother and two younger sisters standing in the driveway near the barn, looking at something in the pasture nearby.

When my mother saw me, she said, "Come here and look at this." I got off my bike and walked over to where they were standing. All our cows were standing close to the barn with their backs to the building, obviously very frightened by something.

Normally all our cows at that time of day would be in one of the far pastures napping, but this day they were on high alert as though being stalked by a dangerous predator.

We had cattle from the time we moved to that farm until I graduated high school, and they almost universally had the same routines every day... Even the tough Angus bull with the nasty disposition was with the other cows behaving as they were. That animal was mean. He would attack any animal or human who dared to venture into his pasture. Seeing the bull cower in fear was extremely out of character.

We looked in the direction the cows seemed to be focused on, and in the tree line at the far edge of this portion of the pasture we saw something unusual. A creature was violently thrashing the young trees. By the amount of vegetation being affected, it was obvious that the creature must have been huge.

My first thought was that the bear we had encountered was back. We had seen no sign of it since the previous year, but it could have come back for more honey. It seemed odd that the bear who had ignored our cattle in the past would behave differently this time.

I have never seen cows react in that manner, especially not our mean bull. It had not been afraid of the bear before, so I wondered what could scare the cows like this if it were not the bear.

I suggested to my mother that maybe the Rock Quarry Monster was down there in the trees. She replied that there was no such thing, and it was probably the bear. I went down to that place the following day to take a look. I wanted to see if there were any bear tracks thinking that if the bear was back, maybe we would have another exciting hunt like the previous year.

The land we lived on was mostly river sand, with some open places. The ground was covered mostly with forest debris and moss. The area was not very conducive for leaving footprints, and I found none. I still was inclined to think the bear had returned, but did not know for certain.

I did not think further about this occurrence or the boy who claimed he had seen the Rock Quarry Monster until a few years later when an event would resurrect those memories.

During the winter of 1970, Dick Whitman was talking with my parents seriously about them purchasing the 40 acres we rented from him. Dick got sick around that time with a very serious infection in one of his legs, which eventually had to be amputated.
Dick never recovered, and he died shortly after the amputation. We were all sad, as he had become a good friend to our family. He had even given me my very first bicycle and always treated us as part of his family. He would be missed a lot by us all.

Dick's wife did not get along well with my father, and she refused to sell my parents the land, so we were forced to move. My parents bought a small farm on ten acres of land in Graham, which was located approximately five miles from Dick's land.

The farm in Graham was above and outside the Puyallup valley and just inside another school district. Many changes were headed our way. I was twelve years old then and thought we had moved a long distance, but the forest was still close to the Puyallup River and near the location where that boy claimed to have encountered The Rock Quarry Monster.

TRACKS IN THE SNOW

I had been allowed to finish out my final year of grade school at Orting Elementary School. I was sad that I would be leaving all my friends I had started kindergarten with and all the places I had come to know so well.

We settled into our new home and I had the summer to get used to it before beginning school in the new district. I was a bit nervous about a whole new school where I knew no one. I made the transition easily enough though. I have always been lucky to make friends easily.

Over the next two years, our circle of friends had become well established. Most of us are still friends to this day. Some lived within bicycle riding distance, but others lived too far away and would have to be driven to our home (or vice versa) to visit outside school.

Our parents did not drive us often, so we would have to improvise if we wanted to do things together on the weekends. If one of us wanted to visit another friend's home, we would get permission to stay the entire weekend. We would then travel home on the school bus with that friend, stay the weekend, and then return to our own home after school the following Monday.

One such weekend was in mid-December 1972. My friend, Mark Barshaw, came to my home to stay the weekend. It had snowed in the middle of the week and was very cold. The temperatures were in the teens.

John Adams, who was another friend and member of our group, lived about a mile from my home. Mark and I decided to go visit John, as there was not much to do at my home that Saturday with snow covering everything. We thought the three of us could find something interesting to do. Normally we would have taken the trail through the forest to John's home. However, it was covered with about three inches of snow and could not be seen.

Mark and I chose to first walk the road to where the railroad tracks crossed and then continue the rest of the distance along the tracks since John's house was near the track line.

The previous summer, the railroad company had brought in a bulldozer and made crude fire access roads on either side of the tracks. At first, we thought we could walk on one of these, but the top of the snow was frozen hard and each step was uncomfortable in the crust. We decided to walk along the railroad tracks themselves, to make the traveling easier.

The distance from where we started on the rail line to John's home was about a quarter mile, and we were about half way to John's house when we spotted something red lying between the rails ahead of us.

When we approached the red object, we found that what we had seen were the intestines of an animal. They were approximately the quantity of what would belong to a medium size dog or coyote. There were a lot of coyotes in that area, so it could have been from one of them, but what was odd was that there were no signs of how these intestines had gotten to this spot.

The quantity was too much for a hawk to have dropped, and trees covered the area. There were absolutely no footprints of any kind in the snow. Only the steel rails were bare of snow, so we would have easily seen any tracks that were there.

(John Adams, his brother Jeff and their father)

This section of the rail line went along the side of a hill. We could clearly see the fire access road below us. That had no markings, so that left only two possibilities. I told Mark to walk ahead and look around while I climbed the 10-foot embankment to our left, and looked around on the fire access road up there.

When I reached the top of the embankment, I was astonished. There were footprints everywhere, and I yelled at Mark to join me. When he caught up, we were shocked at what we saw. In the snow, there were dozens of large human looking footprints, but no shoes. Not only were they over normal proportion in size, but barefoot!

We looked in amazement until the realization hit me that those intestines were not frozen yet, even though the temperature was only about 17 degrees Fahrenheit that morning. Whatever had deposited those intestines between the rails was still very close by. This scared us badly, and we ran down the hill to the rail line and sprinted the remaining distance to John's home.

When we got to John's house, we ran up the front porch and pounded furiously on the front door. John answered the door and we rushed in, excitedly trying to tell fast as we could what we had just found. John had two younger brothers and two sisters, and our excitement caused all of them to be just as agitated.

John's father, hearing all the commotion came in from another room and asked us to tell him what happened. We proceeded to tell him as calmly as we could what we had seen back near the railroad tracks. He got his .45 caliber pistol and camera and said, "Boys, take me to where you found this."

Mark and I led John, alongside his father and siblings back to the place we had found the strange footprints. When we arrived, John's father began taking pictures of the prints.

We found that three individuals had made the footprints. I do not recall the exact measurements of the prints. However, one was larger than the other two at approximately 17 inches in length, while the two smaller sets were around 15 and 16 inches long. We also found a place near the edge of the slope leading down to the railroad line where something very large had appeared to sit down in the snow, with its legs hanging over the edge. Next to the right side of this was a fist impression in the snow. I placed my gloved fist into this—there was easily an inch of room all around my hand.

(Sketch courtesy Jerry Bishop)

We then realized how the intestines had gotten to where they were. Whatever had sat in the snow near the edge of the embankment above the rail line had simply dropped the intestines between the rails from where it sat.

After John's father was finished taking the photographs, he told all of us what he thought had made them. None of us at that time had ever heard the word "Bigfoot" before. He explained that he had seen something on television about one being filmed in Northern California a few years earlier. They were some kind of huge hairy man-like animal that lived in the forests.

In Search of the Unknown

We had started out that day looking for some adventure, and we got a whole lot more than we bargained for! We were fourteen years old then, and normally it does not take much to excite fourteen-year-old boys. We already liked monster movies, but to be told that monsters really existed and roamed the forests near our homes was as exciting as it could get.

We all went back to John's house where we spent the majority of the day all sitting around their dining room table, with hot chocolate excitedly discussing what we found and speculating about the creatures that made the footprints.

At the end of the day, about an hour before dusk came, Mark and I decided we had better return to my home. We did not want to be out after dark knowing what was lurking in the area.

When we arrived at my home, we both told my mother and two younger sisters what we had found that morning. Instead of the excitement and enthusiasm John's father had expressed, my family made fun of us. My mother thought we had too fertile imaginations, and my sisters just made jokes.

We did not know if this would be how anyone else would react to our story, so we decided to keep what we had seen to ourselves from then on. The only people we talked about this find with were very close friends, and only on rare occasions.

I went to John's home many weekends following that incident. We would all sit around their dining room table for many hours discussing the creatures as we had done that first Saturday. Often after talking for a few hours, we would go and search the forest areas nearby, but we never did find any trace of the strange creatures again.

Not finding any more sign of the creatures, we eventually forgot about them and returned to normal activities.

One evening in October 1974 I was going out to feed my second collie, Willie, and as I went outside he was barking at something. He would bark at anything that came near our home. Normally it was small animals or someone coming to visit us. This evening he was facing toward the nearest tree line near our barn and was very excited by something there.

Skunks and raccoons often came into our yard, mostly to eat the cat food left near our back porch for my sister's two cats. My father told me that if any wild animals came into the yard, to shoot them as they might be rabid. So I went into my room and got one of my .22 rifles and a couple bullets.

We kept Willie tethered to his doghouse at night, because he liked to visit the neighboring farms and we did not want him getting into any trouble. I walked over to release him, and he was still barking furiously. I could tell something was in the trees nearby.

When I let him go, I said, "Go get 'em!" He was off like a shot. I never had to tell him to chase anything though. He would chase any animal, dangerous or not.

I followed after him quickly as I did not want him getting bitten by a rabid animal. The tree line was not far from our house, maybe a hundred yards, and he got there quickly.

I was about half way to the tree line and saw Willie acting strangely. Rather than enter the trees, he stood rigid with his ears up as if ready to attack some vicious foe.

(Sketch courtesy Jerry Bishop)

I got within about sixty feet of where he was intently staring at something, when he suddenly wheeled around and ran past me at top speed. I owned him for thirteen years, and had never seen him run away from anything. I was perplexed by his behavior.

He did not stop running until he reached the back steps leading up the porch. He climbed them and sat down shivering visibly. I did not know what to think.
He had never acted this way before.

I walked to the place where he had been standing, and heard something moving in the underbrush inside the trees. I could not see inside the edge of the forest, because the area was covered with blackberry vines and other vegetation.

I put a bullet in the chamber of my rifle, and entered the tree line. It was dusk, but there was still plenty of daylight to see and shoot any raccoon or skunk. I pushed through some low-hanging Douglas fir limbs and walked toward a very large maple tree to where the rustling sounds were coming from.

As I cleared the low-hanging tree boughs I came out into the clearing and got the surprise of my life. Standing in the middle of the clearing under the old maple tree was a huge creature that vaguely resembled a gorilla but was standing like a man.
I was stunned. The only thing in my mind was, "What in the hell is that?!" Then the thought struck me that this must be one of the creatures that made the footprints Mark and I had found a couple years earlier.

Having grown up on a farm helping my father butcher both our domestic and the wild animals he hunted, I could accurately estimate the weight of most creatures by simply observing them. got a very good look at it. This thing had to have weighed around 800 pounds. I was about five foot ten inches tall then, and this creature was a good two feet taller than me.

By this time in the fall all the leafy trees had lost their foliage and this big maple tree had a very thick blanket of yellow leaves at its base. The creature was moving the leaves around in what looked like a leisurely manner with its right foot, to this day I can still clearly see the tendons on the top of its foot to its great toe as it arched its foot upward. The skin that I could see when the hair on its foot would part a little as the foot moved was light colored and looked dirty.

Many people have remarked that in very stressful situations that time seems to slow down, and that is how I recall this encounter. So much seemed to happen, but it must have been in milliseconds. When the creature saw me approach, it stopped what it had been doing and stared at me.

(Sketch courtesy Nathan Reo, except for the hair covering this resembles the creatures I saw)

For what seemed like a long time, we just stared at each other. What scared me so much was not the great size and obvious power of this creature, but the soul-piercing stare it gave me.

(William Jevning and "Willie" 1974)

I have been asked often what this creature looked like. For a quick reference, it looked nearly exactly like the creature Roger Patterson filmed in Northern California in 1967. The slight differences were that the creature standing before me had slightly longer hair all over it. My guess was the hair would have been up to six inches in length but was not uniform length all over. Its shoulders were not as slumped as the creature in Patterson's film. Its shoulders were massive and appeared more squared.

Hair covered most of its face, except its nose and around its eyes. The top of its head resembled a gorilla's quite a lot, being cone shaped but not pointed. I did not see its ears, but they may have been covered by hair.

It had very heavy brow ridges above its eyes, which looked deep set and large, and I think these features gave it a more penetrating stare. Its head seemed to get somewhat larger from the top down toward its jaw, which was very large. I could tell it was able to eat large amounts of food with such a big mouth.

In Search of the Unknown

I couldn't make out its nose very well, but it was flat looking. It had dark brown hair. I honestly paid more attention for those seconds to its eyes. It never gestured with its hands in any way; they just hung there next to its sides. Its arms hung low, lower than people's would.

All of this went through my mind in seconds, and the only sound was from the creature breathing. I had no idea what it was going to do. Obviously my rifle was far too small to do any harm to something so large, so I decided to shoot into the air and possibly scare it away.

I pointed my rifle straight up and fired, and then I heard a noise behind some bushes to my rear and slightly to my right. Suddenly, a second creature appeared.

The second one walked past me toward the first one. It stopped close to the first animal and stared at me the same way that the first one was still doing. This second one looked just like the first one, but it was a head shorter and was not as heavy.

With those frightening, piercing stares, I had no idea what these strange animals intended to do. They certainly did not seem to fear me at all, and I had no real way to defend myself should they attack me. The look they were giving me was not a friendly one.

Without thinking, I took off running for the house, as my dog had done, hoping I was not being pursued.

Recalling the way my family had reacted about the footprints Mark and I had found, I knew telling my parents were out of the question. I called John Adams and quietly told him what had just happened.

We decided to meet at my house the following morning before first light to track the creatures. I called a couple other close friends who lived nearest to me, and they too agreed to meet us at my home early the next morning.

It had been frosty that night, and the ground the next morning was nearly as white as when covered with snow. John and his brothers came before daybreak as we discussed along with another of our friends.

Tracking the creatures was easy. The ground had been soft where I had seen the creatures, and then they left clear tracks farther in the heavy frost. We were able to follow the creatures' movements for more than a mile until the sun came up that clear morning and melted the trail.

We had no clear plan as to what we would do should we encounter those creatures, but were armed heavily enough this time to at least defend ourselves if they acted hostile. We mostly wanted to know more about what we were dealing with, and my friends wanted to get a look at them.

After the trail had melted, we tried to pick it up by searching the softer ground, but we had lost it for good. The trail of prints had gone in a northerly direction toward the railroad tracks where we had originally found footprints.

When we returned to my home, our friend Rick said he should be getting home so we said our goodbyes and all agreed not to tell anyone else what we had been doing. We decided that we needed solid proof of the creatures before saying anything openly.

I went with John and his brothers to their home, where we once again spent the remainder of the day talking about what I had seen the previous evening and what we had seen and done that morning.

Over the following months, we continued to do searches once in a while, but as before, we never saw any signs of the creatures again. We talked about them at times, but kept the subject to ourselves, not allowing anyone outside of those who had seen the footprints to know anything.

One afternoon on the school bus, John and I were having a quiet discussion about a number of topics. We briefly talked about the creatures, looking first to make sure nobody could overhear what we were talking about.

A mutual friend, Jim Dawson, was sitting in a nearby seat and apparently had heard a little of what we were discussing. He quietly joined in asking if he could interview me about the things I had seen. Jim was very quiet and shy and a nice person whom I knew would keep things to himself. I did not see any harm in telling him. He lived relatively close to John, so he could encounter the creatures too. I thought he should know about them.

We decided to go to the high school library, as it would provide a quiet, secluded place to talk. I told Jim of the time Mark and I had found the odd footprints, and of the creatures I had encountered just a few months earlier. He made lots of notes, and then gave me three books that were more like magazines written by a man named John Green.

Green had authored these books about the creatures I had seen. I was astonished that there was so much information about them. The books had a lot of photographs of footprints that looked just like the tracks we had seen. One included a black and white photograph from the film Roger Patterson had taken of a Sasquatch in 1967. To my amazement, it looked just like the two creatures I had seen.

Jim did not say what he wanted the information for, and we did not talk about it again.

This was the winter of 1975. Things went along as normal until later that summer when I would learn what Jim did with the story I told him.

JOHN GREEN & RENÉ DAHINDEN

July that summer was more overcast than usual. For some reason, on one particular day, I was napping in my room instead of being outside as usual. My youngest sister came in and woke me, saying that there were two men here to see me.

I was still groggy and wondering who these men could be. The only men I knew were either my uncles or my father's friends. I slipped on my barn boots and walked out the back door. Our driveway came to the back of our house. We never used the front entrance. As I walked out onto the porch, I was astonished to see René Dahinden walking toward me.

I had poured over the books written by John Green that Jim had given to me, absorbing every detail. I had become well acquainted with the people and information contained in those books, and I knew that Dahinden was the most famous Sasquatch hunter in the world.

I walked toward him, still rubbing the sleep from my eyes, and he came up and shook my hand vigorously, identifying himself and his companion as Dennis Gates. He told me that my friend Jim Dawson had sent the accounts of my finding the footprints and encountering the creatures to John Green. He asked if I would tell him about the things I had seen. I was happy to do so.

We talked for a short time, and then he said that he and Green and a few others were just a few miles north of my home investigating a series of Sasquatch encounters there. He asked if I would like to come to their camp and meet John Green. I said that I would. He was pleased and said that we could talk a lot more then.

René and Dennis then left, and I was once again in shock. First I had encountered some very strange creatures I did not even know existed until just a few years before. Now the world famous people who hunted them had come to my home and invited me to their camp to talk with them.

I did not have a car yet, so I called a few of my friends looking for a ride, but everyone was away from home doing various summer activities. I found one friend home with a car, and he agreed to take me to Puyallup to Green and Dahinden's camp.

As I had mentioned, the weather was very 'un-July' for the Puget Sound region of Washington State that summer. The place they had made camp to conduct their investigations was not far from the only housing development at that time. The Forest Green development had a number of residents there who saw or heard many strange things.

This is the place where the Puyallup Screamer incidents took place, and Green and Dahinden had come to bring their expertise to those events.

As we got to within a mile or two of where Dahinden had told me they were camped, a blinding hailstorm hit us. It came down so hard we could not even see the road. We had to stop to wait for it to pass.

Once we could see again, we proceeded to their camp. We saw a number of vehicles parked along the side of the lonely road. A small travel trailer was parked just off the road in an open area, which I learned later, was a gas pipeline access way.

Dahinden was there and came over to greet us as we parked next to the other vehicles. He said he had been hoping I would be able to visit. We talked more in depth about the things I had seen, and Dennis Gates asked some questions too. After talking awhile, René asked me if I would be interested in staying a few days to help them with the investigation. I said I would very much like to.

My friend Rick and I went to the nearest phone where I called home and asked my mother's permission. She said it would be all right.

I was excited that I was getting to not only meet, but also be involved in an investigation with the most famous Sasquatch hunters in the world.

Dennis had told John Green that I was in the camp. Green told him that he wanted to meet me and talk with me, but it would be later in the day when the local men had left the camp. I did not mind since René Dahinden wanted to talk with me and tell me more about the Sasquatch than was in Green's books.

As we talked, another man approached us, I soon found out he was a Washington State Patrol Officer named Mark Pittenger. Pittenger had his own encounters with Sasquatches and was interested in discussing what I had seen. I soon discovered the scope of encounters in our area was much larger than I had realized.

Apparently, neither I nor any other witnesses in the areas near where I had seen the creatures had spoken publicly about our experiences. The scope of what was being seen is unknown to this day. I am still finding witnesses who also saw those creatures and had never told their stories to anyone, for the same reasons that we kept what we had seen to ourselves.

Pittenger told me that his first encounter with the creatures happened one night while he was parked on the shoulder not far from the camp, near the entrance of what today is the Puyallup extension of Pierce College. He said he parked his patrol car there to do some routine paperwork, leaving his headlights on to remain visible to other motorists.

As he was working, he heard something coming down the rocky slope to the right of his car. He looked up to see what was making the noise and was shocked by what he saw.

He observed a huge hair-covered, grayish-colored man-like creature making its way toward the road. It walked down the slope on two feet like a man and in front of his patrol car, passing within only a few feet, never even looking in his direction. He said the creature took only a couple long strides to cross the road, then climbed the opposite slope and disappeared into the forest.

He said he sat for some time stunned, not having any idea about what he had just witnessed. He said he began researching to see what this might be, and found that other people had reported seeing similar creatures. As he began his own investigation into what was being seen in that area, he found footprints of the very same three creatures we had found footprints of in 1972.

Mark told us that on another occasion he had tracked the three animals through the forest not far from the camp to make a gruesome discovery—one that shook him badly.

(Sketch courtesy Jerry Bishop)

The Sasquatches apparently hunted coyotes, and often emitted screams that imitated them. By following the footprints, Pittenger discovered a place where they had stopped and screamed imitating coyotes and attracted a number of them near. He found that once the unfortunate coyotes got too close, the Sasquatches grabbed them. He said it was obvious the creatures had grasped the coyotes by their hind legs and smashed their heads against alder trees eight feet above the ground.

They had then been torn apart and eaten, leaving a bloody mess and intestines on the ground. Pittenger said he had carried his 30.06 hunting rifle, but this gave him no comfort. He said he crawled on his stomach away from the gruesome scene, hoping he would not draw the creatures' attention.

I had read in one of John Green's books that Indians sometimes called the creatures, "Ventriloquists", because they imitated animal noises. Pittenger was afraid of the creatures from that day, having witnessed what they are capable of.

I told him that we too had found animal intestines that must have come from a coyote or dog of similar size, and he said that was just what he found too. He cautioned us to be careful in the forest and to carry a rifle or shotgun, just in case they were prone to attack a person. I thought that was sound advice coming from a State Trooper. It had substance.

It was fascinating to me to be on a Sasquatch hunt with René Dahinden and John Green. There was no sign of the animosity that came to develop between the two men. I learned soon after of the things that caused the permanent rift between them. I did not know then that the investigation at Puyallup Washington the summer of 1975 was the very last time these two famous Sasquatch hunters would act as a team. They never again worked together.

Thinking back, they did not talk much to each other. There were a lot of people there, and I assumed that both were busy as Dahinden searched during the night hours, while Green talked with local residents during the daytime hours. I thought that was their arrangement to be focused 24 hours a day for the most coverage.

Dahinden asked me if I would work the nights with him and Dennis Gates. I said I would, and it turned out to be quite an experience.

I did get to talk with John Green at length while there. He was quite a lot different to René Dahinden. Green asked me in detail about the things I had seen and experienced, and seemed very interested in what I had to say. He told me of the things that had been reported in the area we were investigating, and said these were probably the same Sasquatches I had seen, since Puyallup was only a few miles north of where I lived.

René Dahinden's approach was very hands on. He and I would quietly walk the roads throughout the night hours. I was used to hunting deer and elk with my father and uncles by this time. I knew to be very quiet and observant—of how experienced hunters did things. I applied all I had been taught by my father to the search with Dahinden. René told me that he really liked my approach.

We searched this area for several nights, sleeping during the daylight hours. Gates would drive to nearby areas and we acted as a team trying to create sort of a pincer movement to attempt to get on two sides of the Sasquatches. We did not have much luck though.

By the end of the third day, René, told me that they had to get going soon. He told me I was, "Pretty good in da bush," adding that he wanted to stay in touch, and thought I would become a very good Sasquatch hunter. He suggested that maybe I could assist him on future hunts. I said I would like that very much.

René told me that shortly before I had come to the camp, he, Pittenger and Dennis Gates, with help from some of the local men, had actually cornered one of the Sasquatches near the backside of what was called Willow Pond . Pittenger had the creature in the sights of his rifle, but for an unknown reason failed to take the shot, and it got away. Dahinden was disgusted, saying that the issue could have been resolved right then and there about the existence of the Sasquatch.

Before Dahinden and Dennis Gates left, John Green came over to where we were talking. Green asked me if I owned a 12-gauge shotgun, I told him that I did. Both he and Dahinden at that time were advocating killing a Sasquatch to once and for all prove they did exist, and proceeded to tell me how to accomplish this if I ever had the opportunity.

They told me to load it with alternating slugs and double ought buckshot. The buckshot was for knock-down power and the slugs to kill. Green said I should aim for the lower spine if a Sasquatch was walking away from me. This would render the creature immobile, allowing me to deliver the killing shot using a slug.

I did not have any opinion of my own regarding killing the creatures back then. They were the experts and I knew they had given the matter a lot of thought.
They told me they had discussed this with an anthropologist at Washington State University, Grover Krantz, who agreed with this method of acquiring a specimen. There had been talk of tranquilizing one of the creatures, but as body weights were difficult to estimate, calculating the dosage might be too far off, resulting in an accidental death, or a drugged creature escaping. The solution would be to simply kill one to prove the existence of the species, and then study it to determine what
would need to be done next.

It all sounded very reasonable then, and I agreed that I would do this if I ever encountered one again.

René and Dennis Gates said their goodbyes and departed for home. John Green too said he had to get back to British Columbia and thanked me for coming and helping them. He asked me to stay in touch and gave me his telephone number and mailing address, (Dahinden had done the same). He asked me if I would be his eyes and ears in the area, and to keep him informed about any new sightings or track finds. I agreed to help him in any way I could.

Green said he had to get his trailer ready to leave. We said our goodbyes and left for home too.

As Rick and I drove to my home where he would drop me off, my mind was into overdrive trying to absorb all I had just encountered and been part of. I was not only officially a Sasquatch hunter, but I was working with the top men in world on the issue. We had been searching for the creatures during the previous three years, but now I had direct access to people who knew the most about them.

After Rick dropped me at my home and left for his own, I called John Adams immediately. He answered and explained that he had been away when I called a few days before, and asked me what was going on. When I excitedly told him whom I met and what happened, he said he wanted to hear everything. I told him I would come right over to his house to tell him.

After cleaning up and changing clothes, I took Willie and walked the trail through the forest to John's home. I was a little nervous from that time on when in forested areas, never knowing when I might again encounter Sasquatches. I had learned so much since the initial find of footprints three years earlier, but the more I seemed to learn, the more questions I had.

When I arrived at John's home, we talked for a long time. He was just as excited as I was and said he wished he had been home so he could have gone with me to Green and Dahinden's camp.

We talked of doing new Sasquatch hunts, but John was unable to take part in many of the adventures yet to come. One incident shortly after Green and Dahinden returned to British Columbia, Canada is worth mentioning, both because it was Sasquatch related and for its humor.

In Search of the Unknown

John and I had talked about keeping an eye on the south hill Puyallup area as Green and Dahinden had asked. Also, John was interested in seeing where we investigated when they were there. Most of our group had their driver's licenses at this time, but few owned cars. Some of us could borrow our parent's cars to go places, and John had borrowed his parent's station wagon so we could visit where Green and Dahinden's camp had been to look around.

John's younger brother, Jeff, came with us on this excursion. He had been part of so many previous searches for the creatures and was also very interested to see this place too.

We arrived well after dark to the place where John Green's trailer had been parked. The three of us decided to walk the tree line along the gas pipeline access way.

We decided to behave as Dahinden had, walking quietly, not using our flash lights, and listening for any animal movements. Once in a while we would turn on one of the lights to see if there were any animal eye's shining nearby. This would only be done for a few moments as we did not want to disturb the creatures and drive them away.

We would walk for a few minutes, and then stop and John would take a quick look ahead of us while I would shine my light behind us. During one such stop, I was moving my light beam along the opposite tree line, and at one time caught eye shine just inside the trees facing our location.

At first I did not say anything, and casually moved the light back across the same spot to verify that I had indeed seen something. When I shined the light again at the place I had seen eye shine, there were indeed two eyes staring at us. I could even see them blink a couple of times. I turned off my light without reacting, then quietly I told John and Jeff what I had just seen. They did not look like deer eyes and were way too high above the ground to belong to a deer.

Though we were very nervous, I said we should act like nothing was out of the ordinary and focus on the place where I saw the eyes. We could then shine the flashlights on the spot and possibly get a better look at what might be there.

We got into position and turned our lights on at the same moment. All hell broke loose! We saw eye shine briefly, and then something very large took off running across our front. We could easily tell whatever was just inside those trees was running on two feet. It took very long steps as it ran from our right to our left in the direction we had been going.

We cannot say for certain it was a Sasquatch as we did not see the creature, but we had all seen and heard deer many times, and this for certain was no deer. It ran for about a dozen very long "smash...... smash....... smash" steps and then it stopped. That's when the three of us got scared; because we did not have any idea what it would do next and we were a long way from the safety of the car.

We were afraid it would come back, especially after what Pittenger had told me about the smashed coyotes. I whispered to the others to stay calm and to walk casually back to the car. That worked for just a few steps, then terror took over and all three of us bolted at top speed!

John unlocked his door, jumped in behind the steering wheel, and unlocked my door as quickly as he could while trying to start the engine. I was saying, "Let's go! Let's go!" Both of us completely forgetting poor Jeff who was still locked out of the vehicle! Jeff was pounding on the side of the car and screaming to be let in as John started driving away with Jeff hanging onto the door handle.

John only went a few feet then stopped. We unlocked Jeff's door, he jumped in and we sped away. I guess we weren't quite ready for night Sasquatch hunting yet, but that story still makes us laugh today. Was a Sasquatch there? We have no idea, as we never returned in the daytime to see if there were any tracks to show one had been there. However, given the ongoing events that were happening in that location, it is very likely one was there.

By September, following my meeting with Dahinden and Green, I began receiving correspondence from both men. I especially enjoyed getting mail from René as he always included something in his letters. (Often postcards he had made with the famous frame from Patterson's film showing the Sasquatch turn to face Patterson, or newspaper clippings of sightings, or other interesting bits of information.)

Up to this point I was under the impression that John Green was the primary investigator and René was helping him, so I would send Green anything I came across I thought he could use.

René wrote that fall inviting me on an expedition into the Garibaldi wilderness in British Columbia. The expedition would be by float plane, and would comprise a number of legs into areas that were only reachable by air.

I immediately asked my parents, and was mortified when I was told I could not go. Their reasoning was they did not know Dahinden well enough yet. I can see their point now, but at the time it was a giant disappointment. Years later René told me that it was probably just as well I did not go along as the float plane nearly crashed. We both laughed as he recounted that trip and its lighter moments, but in retrospect it was probably better that I did not go.

By the early part of 1976, we began talking among our group of friends seriously about how to begin the searches John Green and René Dahinden had asked me to conduct. We thought that we should act as an organized effort, and I had already come up with a name for our group shortly after meeting Green and Dahinden.

In September 1975, I dubbed our new group the Pierce County Sasquatch Investigation Team. The name reflected the geographical area we were working and the purpose of our group.

By January of 1976, we had all decided that this was the direction we would go, and we began talking about how we should begin as an organized group.

We were talking one day in school during some free time, and I asked if anyone had heard of any strange things happening in their areas? We all lived far enough apart that we covered the area from Graham out to the town of Roy on the opposite side of Fort Lewis.

No one had heard anything except Paul, who said there had been some weird screams out at the Clark Ranch recently. I said he should ask permission from the owners for us to have a look around on their property since he lived close to them and knew the family.

The Roy area has had a history of these strange creatures that still continues to this day. Paul said there had been strange things happening even at his home which was located in Roy. This might seem improbable, but even today Roy has a small population and few buildings.

Paul was often prone to embellish the facts, so the rest of us took what he told us with a grain of salt. Not long before he told us of the stories from the Clark ranch, he told us that he had been hearing strange noises around his family home at night as if someone was walking on the wood walkway near the back door.

We did not think a Sasquatch was sneaking around their house, and did not give what he said much weight. A few weeks later though he called me and said their rabbit cages on his property had been torn apart and the rabbits either killed or scared to death. I have always been the type of person to believe that when a claim has been made there must be some evidence to support it. I wanted to see for myself the cages and if there was some reasonable explanation for what had happened. I did not have much hope that something out of the ordinary had gotten to the rabbits.

Milo Rogers, myself, Scott Martin, Kirk Goddard and Brian Godwin went to his home the weekend following the incident to take a look. Paul's mother had originally told him that dogs had probably killed the rabbits, but after we examined the cages we came to a very different conclusion. The cages were made of sturdy wood two by six inch boards.

The wire mesh was the heaviest gauge steel used for animal pens. It was galvanized and very tough. The arrangement was constructed in an 'L' shape and built as a single piece that was attached to the backside of the house. It was very sturdy.

There were about ten individual cages in all, and the bottom of the structure measured 18 inches above the ground. Below each cage was a flat pan used to collect the rabbit droppings that fell through the mesh that made up the cage bottoms, sides and tops.

The first thing we saw when we entered the back of the house was that rabbit fur covered a large area of the yard and some of the cages had been literally torn open. We saw that several of the remaining rabbits were untouched in their cages, but they had died from unapparent causes.

The cages that had been torn open were done so from the underside. For a dog to have torn the cages open that way, they would have had to lie on its back with feet straight up, which would have been a difficult feat for any dog. Plus the rabbit feces in the collection pans were completely undisturbed.

In addition to the extremely awkward position, lack of space and nothing under the cages being disturbed, the steel mesh was the telling evidence. In several places the mesh had holes punctured through it, which were approximately the same size. They measured between one and two inches across, and all were near one of the front corners of the bottom of the cages.

The appearance of the torn open cages gave the strong impression that a big hand had poked large fingers through the steel mesh, and clenched the hand, crushing the mesh near a corner. Then, a powerful arm pulled the bottom of the cage out tearing the mesh along two sides of the wood framework.

The rabbits missing from these cages were nowhere to be found, except for fur scattered across the yard near the cages. There was no other corroborating evidence such as footprints to indicate what had done this, but the ground was very hard near the cages and was not conducive for any footprints.

We could not prove that a Sasquatch had been responsible for tearing open these heavily reinforced cages, but we did not have any other logical suspects.

We left that day with nothing more than unanswered questions, but there had been Sasquatch activity reported in the general area—even by soldiers at Fort Lewis nearby.

At this point anything seemed possible.

THE CLARK RANCH

Once we obtained permission from the Clark family to camp and search their property, we decided who would be part of the expedition and when we would conduct it.

Four of us were able to go on this first trip, myself, Milo, Paul and Bob. We chose a weekend in February or March that year and would take the school bus to Paul's house in Roy. Once we arrived at Paul's home, we took our packs that we had brought with us and stored away any last-minute items we decided we might need.

We had not brought a lot of things: a change of clothes and food for a couple days. We brought along an old heavy canvas cabin tent that Paul had. It took two of us to carry it due to its weight and size. We took turns carrying the tent in pairs as we hiked to the Clark ranch.

We had decided to walk along the railroad tracks that led out of town and passed by the Clark's home.

Once we arrived at the ranch, we decided a rest was in order after hauling the heavy tent for a mile or so. We also wanted to check in with the Clark family and find out if any new noises had been heard there.

They came out to talk with us, and said they heard the strange screams almost nightly. They said we were welcome to go anywhere we liked on their land, but they would not join us. They pointed to where the odd noises came from and we thanked them and moved out in that direction.

We hiked across open fields to the nearest tree line, where we entered the forest and kept going in the direction they said the screams came from.

We were ill prepared for searching from the word go. I was the only one of our group that had a camera but had not brought it with me. We had only one flashlight among the four of us, and we didn't really have any plan. We were mostly going to see if the stories had any substance. I am not sure what we thought we would accomplish if we did encounter a Sasquatch. In those early days, we had not learned to plan for contingencies, but we went ahead anyway.

We hiked cross-country, having no trails to navigate by. Soon we came to a stand of cedar trees with a nice grassy area in the middle that looked like a good place to make camp.

We were in close proximity to the Nisqually River, which seemed as good a place as any to set up our base of operations. We figured if anything were out here, this would be the best area to look.
We set up the tent near the edge of the open area and collected some large rocks to make a fire ring. We gathered a pile of dry wood and then made a roaring fire, as it was getting late in the day and was starting to get cold.

The camp had become cozy enough now that the tent was set up with our packs stored inside it, and a roaring bonfire heating up the area. It got dark quickly that time of year, and we decided that we would cook our dinner and discuss how we would begin conducting our search of the area the following morning.

In Search of the Unknown

After eating the canned Beanie Weenies we had brought with us, we sat back near the fire to talk. We did not even have any camp chairs, but were comfortable leaning against a log near the tent or just kneeling by the fire.

It was nearly 10:00 p.m. We had been talking quietly for several hours around the fire about all sorts of things. Suddenly, from just a couple hundred yards away, something big let out the ear-piercing scream. All four of us jumped at once. This was no animal we knew. Shortly after this scream, we heard another answer from much farther away.

This call and answer went on for about a half an hour. We listened intently to see if they were getting closer to us. The callers did not seem to be moving, and now had stopped.

We listened on edge for what seemed like a long time; not even talking among ourselves for fear that would put us all in danger. Paul had gotten sleepy and wanted to go into the tent to sleep.

I suggested that we work in teams of two, with two of us awake at all times in case any uninvited visitors came into the camp. Everyone agreed. Milo and I were one team and Paul and Bob made the other.

Milo and I decided we would take the first watch while the other two got a couple of hours of sleep, then we would switch.

Bob said he could not sleep because the screams had made him too nervous. I told Paul to go ahead and get some sleep, and we would figure out the guard-duty arrangements.

Paul went into the tent. The three of us sat quietly by the fire talking in low tones about the screaming we had heard. After about twenty minutes, we heard Paul rustling around inside the tent and wondered what he was doing.

(Sketch courtesy Jerry Bishop)

We had begun joking about Paul's activity when he bolted out of the tent, nearly knocking it down in the process.

He had a wild-eyed look on his face, as though he had been scared badly by something. He shouted that trying to pull him out of the tent wasn't very damned funny. I said, "Well, that's a pretty funny thing to do, but none of us had thought to do that."
I asked him if he was sure he hadn't dreamt it. He said that he was not able to get to sleep and was in his sleeping bag listening to our conversation.

I asked him to tell us what exactly happened. He said at first he had heard some noise at the side of the tent. He thought one of us was trying to get something from one of the packs, and reached under the side of the tent instead of going inside and disturbing him.

That old cabin tent did not have a floor in it as modern tents do, so it was easy to reach under the edge to the inside. After this went on for a few moments, he said that a hand grabbed the backside of his sleeping bag. Paul said he had been lying on his stomach, and this hand had grasped him across his lower back.

I asked him how he could tell it was a hand. He said he could feel it. I asked him if he could feel from the heel of the hand to the fingertips. He said yes. I then asked him to show us where the fingertips started on his back and where the heel of this hand had been. Paul was a big guy, and when he showed us where the hand had been, it was huge. We estimated that this hand must have been more than a foot in length, possibly more. Paul said that this hand grasped him and began to pull him out of the tent; that's when he began to struggle.

Paul said that when he began to fight, the hand released him. Apparently the visitor had not realized it had grasped a living creature instead of the food we had cooked for dinner.

I was still not convinced, and asked one of the guys to toss the flashlight to me. I went behind the tent to where he said this intruder had been. I said that if he were not dreaming, then there must be some signs of someone or something there.

To my amazement when I went behind the tent, there were large footprints like those I had seen before in the soft soil. The visitor had approached the tent from the thick woods behind the tent, leaving a trail of footprints, and a place where it had knelt down not very far behind the place I was kneeling at the corner of the tent.

None of us had heard any noise, and it had gotten very close to us. Only Paul's encounter had alerted us to its presence, otherwise we would have never known it was there.

(Sketch courtesy Jerry Bishop)

As we all looked at the footprints, we were in awe. Something so large had gotten so close to us without our detecting its presence.

We were now on high alert and began to build the fire larger in hopes it might be a deterrent to it coming too close.

About this time the screams began again. They were similar to what had heard earlier, with one being fairly close by and the second much farther away.
There were many tree frogs in this area, and it was a little marshy this close to the river. It sounded like thousands of frogs all croaking at once. Suddenly, all of them to one side of us stopped at once.

Anyone who has ever experienced this will know it gives one a very eerie feeling. You know something is out there causing the frogs to become silent.

(Sketch courtesy Jerry Bishop)

Tree frogs always stop croaking when a person or animal gets near them, so this was a good way to detect when we had an intruder nearby.

After a few minutes, the frogs began to croak once again all around the camp. We did not hear anything more for a while and began to relax and settle next to the fire and discuss the things that had happened. None of us was sleepy now. We knew we had better remain on guard throughout the night.

We began to think our visitors had gone away when once again to one side of the camp, all the frogs suddenly became silent. Then, moving around our camp in a counter-clockwise direction, the frogs stopped croaking on one side of camp, then the other.

We could easily tell we were being circled by something. The frogs gave the intruders' movements away. This went on for several hours, and we began to get more frightened as the night went on. We were too far away to pack up camp and leave, not to mention having one flashlight that badly needed fresh batteries.

We also had come cross-country and would need daylight to see our way out, so we had to sit tight and wait the night out there.

At one point we knew that two of the creatures were circling our camp. As before, each time they came close, the tree frogs ceased croaking. We always knew which side of camp they approached as the frogs stopped in just that area.

They stopped on one side of camp, then moments later, the opposite side. We believed there were two of them stalking us. At the same time the screamers began again, but now they were both much closer.

We decided that we needed to keep the fire a respectable size throughout the night until daylight and then we could leave. We were afraid to go too far from the fire. Unfortunately, after a while, we began to run out of the near supply of fuel.

Throughout the night, we experienced sessions of long quiet, then terrifying moments when as they looked us over. Then they would leave again. This pattern repeated a number of times.

Once, while we had decided to stand close to the fire and face all four directions to be on maximum guard, Milo had turned to tell me something. He was facing the tent and was between the door of the tent and the fire. When he finished speaking to me, he returned to once again face the tent. This time one of the Sasquatches was standing right behind the tent in open view in the bright firelight.

(Sketch courtesy Jerry Bishop)

Milo literally jumped and yelled, "Holy shit!" The creature had been glaring right at him, and the commotion caused by his outburst had allowed the creature to step silently back into the shadows.

By around 4:30 a.m. we began to run out of firewood and were too scared to venture farther beyond the firelight for more. We all became very sleepy too. We decided that all we could do was to sit down shoulder to shoulder with our backs against a large spruce log. We thought that if we fell asleep with our machetes in hand, any attacker would us simultaneously, and as a group we could fight it. It was a naive thought, but we were young, inexperienced, exhausted, and just didn't care anymore.

We all sat down together and proceeded to fall asleep. Apparently we had become boring to our visitors and they left the area. We awoke at first light to find we were all still there in one piece and unharmed. We hurriedly packed things up and left the area as quickly as we could.

We did not stop at the Clark home to tell them we were leaving, plus it was very early and we did not want to wake them, or tell anyone what we experienced the night before.

When we arrived at Paul's house, his mother was awake and made us a huge breakfast as we told her of the night's events.

I don't think she believed us, but knew we were all pretty upset and she just listened. She then offered to take each of us home instead of our waiting until Monday to return on the school bus.

I never said anything at home. We kept all our experiences to ourselves. I wrote a letter right away to John Green telling him all that had happened and sent it Monday morning.

About two weeks later, after I had gotten home from school one day, John Green called and said he was near Seattle and would be at my home in approximately an hour. He asked if I would take him to the Clark ranch, and I responded that I certainly would.

I thought that if anyone could do something about the Clark ranch events, it was Green. I called my friend Paul and told him that Green was on his way and asked him to call the Clarks to ask if it was all right if we come out there. I then called Scott Martin and asked him if he wanted to come along. He said he did and would be right over.

Paul called back and said the Clark family told him it would be okay if we visited, and I said we would be there as soon as Green arrived. Before too long, Scott arrived at my home, and Green showed up shortly after.

We got into Green's Volkswagen van and he drove us to Roy, where we picked Paul up. When we arrived at the Clark ranch, it was beginning to get dark. As soon as Green parked the van and we got out to be greeted by the Clark boys, the screams began.

We all stood there motionless for what seemed like a long time listening to the screams. Green then said that he could not stay long and had to get on the road back to British Columbia.

We all got back in his van, and he dropped each of us off at our homes. I thought it was odd that he did not even record the screams, but I did not say so until I wrote a letter to René Dahinden telling him of the camping trip and Green's subsequent visit.

That's when Dahinden began telling me that he didn't consider Green to be a real Sasquatch hunter, and that John did not know what he was doing. Years later when I would visit Green at his home in Harrison Hot Springs, he would always say he was still kicking himself for not bringing his recorder that night.

I did wonder though why he never went back with recording equipment. So I began to have my doubts about John Green and his intentions about proving the existence of the Sasquatch. I was beginning to see the major differences between him and René Dahinden.

I did not side with one man or the other, and did not think much about their differences or want to be involved in them. We continued to search places we thought we might find evidence of the Sasquatch, but with little success.

(From left, my sister Rosemary, my mother and myself at our home the summer of 1976)

The summer of 1976 was our last summer in high school, Milo had a summer job and the rest of us were busy with other activities. We did little that summer in regard to the Sasquatch.

The fall of that year was the last big adventure we would have as a group.

MOUNT SAINT HELENS EXPEDITION

During late September or early October 1976, I received a telephone call from a college student in Portland, Oregon. He introduced himself and told me that he was writing a paper about the Sasquatch. He had spoken with John Green who recommended he call me, as I had actually seen two of the creatures. Green had told him I would be a good source of first-hand information for his paper, and had given him my contact details.

He asked me if I would tell him what I had seen, and I explained how we had first found footprints in 1972, and had never heard the word "Bigfoot" until then. I went on to tell him about my encounter with two of the creatures, and that my friends and I had begun searching for them to help John Green and René Dahinden.

After I finished telling him what I experienced, he asked me if I had ever been to the site near Mt. St. Helens where the miners had been attacked in 1924. I told him that I had not been to that location, but had hunted deer and elk with my father and uncle on the western side of the mountain.

He recommended that I should go there if I ever got a chance. He said that he and a few of his friends had gone there recently and he gave me directions to the place where the miner's cabin had been located. He thanked me for the information, and I thanked him for the directions and wished him well with his paper.

I knew miner, Fred Beck's story of the 1924 attack from one of John Green's books and thought a trip there would be interesting. I knew that hikers still occasionally saw Sasquatches in that area, and thought it would be a very likely place where we might see one too.

At school the following day, I told my friends about the phone call, and suggested we should make our own expedition to the site.

That fall of 1976 was the beginning of our senior year of high school. Most of our group of friends had already completed the required courses to graduate, so we could afford to take a few days from school to make such a trip. We began to make plans to go to Mt. St. Helens.

I was the only member of our group who had ever been to that mountain. I had only been on hunting trips and did not know the area very well. We did our best to plan our route and other details of the trip. We did not wish to repeat the Clark ranch episode, so we decided to bring some weapons.

We knew better than to bring any firearms, even though several of us had numerous rifles. In those days, it was just something we knew we should not do except during hunting season. We felt we could take care of ourselves other ways.

We each owned a sling shot called a Wrist Rocket and we made solid shot projectiles. For the unfamiliar, imagine a piece of pipe, but instead of it being hollow stock it is solid. We cut this into one half-inch sections—thank goodness for metal shop in school!

We were pretty good shots with these and felt they were enough to keep any threat at bay.

We got our backpacks prepared to stay several days on the mountain, except for food, which we decided to get along the way to Mt. St. Helens.

On the morning of November 10th, we were all waiting in front of our high school with our packs. Our friend Scott Martin had borrowed his father's new Volkswagen van for the trip and picked the rest of us up after an appointment he had that morning.

In Search of the Unknown

Six of us decided to make this adventure: me, Scott Martin, Milo Rogers, Brian Godwin, Kirk Goddard and Paul Willis. This would turn out to be the last of our adventures together as high school students, but one of the most memorable ones.

On our way to Southwest Washington, Scott drove while I sat in front to navigate. Milo and Paul sat in the center seats, with Brian and Kirk in the rear of the van.
We had only been driving about a half hour when the first incident happened.

Paul was the member of our group who seemed to have more than his share of mishaps. Invariably, when we hiked somewhere, something always fell off his backpack. His standard way of notifying the rest of the group that we had to stop and wait for him to recover the item was by saying, "I dropped my gear."

Paul loved old military equipment, and on this trip he had brought along some very old military C-rations. He claimed that his uncle had brought the rations home from the Korean War. By the looks of them he may have been right.

Eating was always high on Paul's priority list, and as we drove south on Interstate 5, he decided he needed a snack. He chose a can marked 'beans and franks', or something close to that. When he opened the can, he got more than he bargained for.

He tapped me on my shoulder, and asked me if I thought the contents of the can were fit for human consumption. When I turned around I saw the beans and franks were actually coming up out of the can under their own power!

I am sure this movement was caused by some kind of gas release, (no pun intended!) from the beans being in the can so long. Not only did they look inedible, but also the fact they were moving unnerved me. I told him I thought he should get rid of them before we found them wanting to eat us.

I turned back around to watch the road ahead and talk with Scott, paying no more attention to Paul's dilemma. Paul then, instead of depositing the suspect can into a trash bag, opened the side window of the van and was about to throw it out onto the freeway.

Scott looked in his rear view mirror in time to see what Paul was about to do, and yelled loudly, "NO! There's a state trooper right behind us!" Milo, Brian and Kirk simultaneously grabbed Paul's arms preventing him from throwing the can out in front of a Washington State Patrol Officer.

Milo asked him what the hell he thought he was doing. He then told him, "Just put the can into the trash bag." Paul said he thought it wasn't safe to have the can in the van, I said we could get rid of it when we stopped next but that it was fine for now.

The rest of the drive was without incident, until we reached the small town of Castle Rock. Castle Rock was the point where we would leave Interstate 5 and drive toward Mt. St. Helens, and it was the last place we would be able to get food for the trip.

We stopped at the local grocery store, where we divided into two three-man groups, each with a shopping cart. We had absolutely no plan as to what we should take with us, or even a consideration for what would fit into our backpacks.

Each of us just grabbed whatever we fancied and put it in the cart. When both groups returned to the van, we had far more food than we needed, and nothing that even made any sense for this sort of venture.

Looking back, it was very humorous: Here six of us had two shopping carts full of food and not one of us even commented on the quantity, let alone how we thought we might transport it anywhere.

We loaded all the provisions into the van, got in, and drove on toward Toutle. Scott mentioned that we would need to fuel up the van soon. We came across a lone gas station. While Scott filled the tank, the rest of us went into the adjacent small store and purchased snacks. I wonder now with so much food that we had just bought why we thought we needed more snacks.

When Scott finished filling the tank, we got in and drove on toward Spirit Lake. It was getting late and we would have to find a place to camp for the night and resume our journey to the site of the miner's attack the following day.

We soon saw a sign that read Spirit Lake Ranger Station, and we could see Spirit Lake itself. I was happy that we had found this place since the trail, according to the directions I had, would be close by.

We decided that since we were nearing the trail we could get a good night's rest before beginning the hike the next morning. There was a large blacktopped area across the road from Spirit Lake, so we drove in there to see if there was a place to camp.

There was no campground. The area looked like a new parking lot. We drove to the far side of the area, which were easily several acres. We decided on a flat spot we could put up the tent just off the blacktop and began to gather wood for a fire, as this was our standard practice when setting up a camp.

The moon was full that night and very bright, so much so that we did not need to use flashlights. We soon had a big problem. The wind was too strong to get a fire started. Even using some white gas for the lanterns proved ineffective. We gave up the idea of making a fire that night.

We had decided to delay putting up the tent until the wind died down. None of us was sleepy anyway, so we decided to have a look around the area.

The moon was so bright that it was almost as easy to see as in daytime. The entire area was made up of light gray pumice, which reflected the light and added to the visibility.

We found a well-used deer trail that went on top of a small ridge and decided to follow that to see where it went. The pumice was slippery to walk on, and we soon discovered that we could slide down the slopes of the hill as if it were covered with snow.

We found a flat place where we could sit and slide down the hill for about a hundred yards or so. It was a lot of fun.

Sometimes small fir trees at the bottom of the slope would stop one or the other of us, which was not the best way to stop! We were not thinking about anything except having fun. After a while of sliding down then climbing back up to the starting point, we noticed our backsides getting numb.

In Search of the Unknown

It had been a chilly November night, but not really cold. We discovered that the backsides of our jeans had literally been sanded away from repeatedly sliding down the pumice hill! With our rears hanging out in the breeze, we decided that we had better return to the van and change clothes.

When we reached the van, Scott said he could not find his wallet or the keys to the van. This was not good. Obviously we needed the keys for the van, but we had also pooled our money and given it to Scott for gas on the return trip. We went back to where we had been sliding down the hill to look for the wallet and keys, which must have fallen out of Scott's pocket there.

This could have been an impossible task finding the lost items. Since the pumice was soft and a couple of feet thick. The wallet and keys could have easily been buried.

We caught a lucky break and found the wallet and keys on the flat place at the top of the hill where we began our downward slide. As we found the items, Kirk yelled to the rest of us that he had found something else.

We all went over to where he and Brian had been looking for the wallet and keys, and Kirk said, "I think I found footprints." He had indeed found a long trail of Sasquatch footprints going up the top of the ridge.

The tracks measured approximately seventeen inches long. The gap between each consecutive footprint was farther than any of us could stretch our legs to match. In the bright moonlight, we could see hundreds of tracks in a line extending up the slope as far as we could see.

Scott retrieved the camera he had borrowed from his parents and took some photographs of the footprints. In the soft pumice, the prints were not well detailed, but it was obvious what had made them.

We were not about to follow the line of tracks. It was late in the night and with our prior encounter at the Clark ranch, we were not willing to repeat that experience any time soon.

We had no way of knowing when those footprints had been made, but they were very easy to see. When we first went to the top of the pumice hill, we had not seen them. We have speculated since that perhaps the creature was in the area and heard the noise we were making whilst laughing and sliding down the hill and came to see what was happening.

We even thought the wallet and keys looked as though they had been placed carefully at the place we had all begun our slide down the hill. We speculated that the Sasquatch had found the wallet and keys after we returned to the van to change our clothes and placed them on the small flat area so we would find them.

I don't know why we speculated this. Maybe we just let our imaginations go too far, being more than one hundred miles from home in the middle of the night on that lonely windswept hillside with that long line of Sasquatch tracks.

What likely happened as Scott sat down before beginning one of his slides down that hill the wallet and keys just fell out of his pocket there. Nevertheless, it was odd to see all those Sasquatch footprints so close to where we had been enjoying ourselves.

Since we were not going to follow that line of tracks, we decided to go back to the van, change our pants and get some rest so we could start the next day journey to the site of the miner's attack.
Once we arrived back at the van, the wind was still blowing too strongly to build a fire. We decided to just attempt to sleep in our seats in the van.

Sleeping that way made for an uncomfortable night. The next morning, we awoke excited that we had found Sasquatch tracks on our first day on Mount St. Helens. We were excited by the possibility that we would find more evidence, and the trip had been worth the effort.

At some point during the middle of the night, Kirk and Paul needed to get out of the van to relieve themselves. None of us were able to sleep very well, so this was a good opportunity to mess with Paul and Kirk. We were always on watch for any chance to play some kind of joke on one another. When they returned, they found we had locked them out of the van.

As they were trying to get us to unlock the doors, an unusual thing happened. Being well in the month of November there was no one else at Spirit Lake and no cars in that large parking area except for ours. A lone car then drove in the far side of the blacktopped area and began doing doughnuts, (turning tight circles with the car while trying to go as fast as they could).

The driver of that car was probably a teenager just like us out having some fun, but in the process was creating a huge oily looking cloud of burned tire smoke. The cloud was amazingly large, and with the wind blowing as it was, made the cloud stay close to the ground and head directly toward us.

The car then drove out of the parking area and away. Paul and Kirk were now pleading as if for their lives to be allowed in the van.

We thought this was really funny now, and the more we laughed, the more they pleaded. As the cloud of smoke got closer, Brian and Scott taunted the two of them by saying, "You're both gonna die!" I thought we should have let the guys in, but it was funny to watch and all they had to do was hold their breath for a few moments until the cloud passed.

When the cloud overtook us, I was amazed how thick it was and that we were unable to see even a fraction of an inch through it from the windows. We could hear Paul coughing loudly, but the cloud soon passed and it was clear once again.

When the smoke had passed, we unlocked the doors, Kirk seemed fine and called us all, "Butt holes", but then he laughed and said he would have done the same thing. Kirk had asthma, and I asked him why he was not coughing as Paul was. He said he just held his breath until the cloud of smoke cleared. Brian had asked Paul why he didn't hold his breath too. Paul just smiled and sheepishly shrugged his shoulders as if to say, "I don't know" and this made us all laugh that much more.

Fortunately the rest of the night passed quickly and uneventfully. We managed to get some sleep.

When it was daybreak, we woke and got underway. We were anxious to reach our destination and set up camp.

We had been right on track navigating and were much closer to the trail we needed to take than expected. We soon reached the end of the highway and the beginning of the trail to Windy Pass. (Today, after the 1980 eruption of Mt. St. Helens, the area is called Windy Ridge, but then it was called Windy Pass.)

We parked the van at the trail head and donned our backpacks. That was when we realized we had too much food to carry in one trip. I said we should take as much as we could, and find the place the cabin once had been. There we would establish camp, and then return for the tent and remainder of our supplies.

In Search of the Unknown

We all agreed and then set off up the trail. Brian and I were in the best physical condition, so we led the party toward the pass. Scott and Milo followed, with Paul and Kirk bringing up the rear.

The pass was a long way up the northern slope of the mountain, and the hike was difficult. We often sunk knee deep in the soft pumice. The hike to the pass took a long time, and we stopped often to rest briefly before pushing on.

Paul was always a source of entertainment for the rest of us, and we never knew what to expect from him. All of us had dressed appropriately for the hike, wearing jeans and leather boots. We wore light jackets and hats in case of rain.

Paul, on the other hand, was a sight all unto himself, as he was dressed a bit differently. Paul had also worn jeans, but his green rubber boots were designed for farm work and made walking difficult. His feet slipped in them often.

Paul also had on a long trench coat that came to the tops of his boots. To top off the ensemble, he wore an old steel army helmet. He was quite a sight!

None of us made any comments about his appearance. We thought that if this were the way he wanted to dress, he would have to deal with the consequences.

By mid-day Brian and I reached the narrow gap in the rock that made Windy Pass. We decided to rest while the others caught up.

Past this gap in the rock, was a wide plain area named the Plains of Abraham. This is a shelf that is on the eastern side of Mt. St. Helens. It survived the eruption of 1980.

Brian and I had walked about twenty minutes ahead of the rest of the group, so we had some time to relax and talk.

About ten minutes into our discussion about where we would set up camp before returning to the van for the rest of our provisions, Milo came rushing into the gap where we were sitting.

Milo tried talking while catching his breath, and all we could understand was something about Paul and, "Mum, mum, mum dropped my gear" Brian told him to slow down and speak English! Milo took a couple of deep breaths and said, "Dumb shit dropped his entire backpack!"

Brian and I exchanged a look of exasperation. I asked Milo if he told Paul to pick up his pack. He got an expression on his face which suggested he had just had a brilliant idea. He held up his index finger and said, "I will be right back!" Then he ran off back down the slope toward the others.

After Milo had left, Brian looked at me and said, "Paul is a dumb ass." Milo returned after a short time, more excited and angry than he had been the first time. He again was trying to breathlessly to get his words out. We told him to sit down and relax and tell us what happened.

When he was able, Milo explained that when he returned to where Paul had dropped his pack, that instead of trying to pick it up he had stood there and watched it roll down the hill.

As the pack descended the slope, it came open scattering its contents all over the side of the mountain. I just took a deep breath and said that since we had to make a return trip anyway, we would collect the pack and its contents at that time.

Milo stayed with Brian and I until the rest of the party arrived at the pass. We told them to rest up before we resumed our journey.

When we were ready, we exited the rock passage that led out onto the Plains of Abraham. There we encountered a group of men on horseback. We introduced ourselves. They said they were hunting Elk. We explained that we were hiking and would be camping for a few days.

They invited us to visit their camp if we happened to hike farther south on the plains, but we had no idea if we would be close enough to them. However, we accepted their invitation and they went on with their hunt.

We still had several miles to go to reach the place where the miner's cabin had been and the upper end of Ape Canyon.

A couple of weeks after the end of our expedition we heard that a group of hunters had died of exposure during the time we were in that area. I often have wondered if the group we met was the same hunters. I thought it was very sad if they were, and marveled at how lucky we were not to have suffered the same fate being so unorganized and ill prepared.

We hiked for a few miles until we came to a stand of trees near the eastern edge of the plains. Beyond the edge was a sheer drop off into an abyss called The Dark Divide. We did not know this back then, but the view of Mt. Adams was beautiful. We loved the area.

We came to and entered a stand of trees, and found that this was the place we were looking for. The college student I had spoken with told me certain things they had made there to mark the location and we easily found what he mentioned.

We found a place to set up camp and decided that the three of us who were in the best physical condition would make the return hike to the van while the other three would begin making camp and a fire.

Brian, Milo and I were the three to make the return trip. We set off immediately so we would be back before nightfall. We emptied our packs and set off for the van. The plains were beautiful and easy to hike through. They were covered with sparse grass and mosses, with the trail easily visible but not well used.

The view from the pass was breathtaking. We had not realized just how far we had gone until we looked at the van from this vantage point.

The van looked no larger than the head of a pin from the pass, and we realized we had our work cut out for us getting the tent and food back up here.

As we descended the slope, we passed Paul's backpack and saw the contents scattered about but paid little attention to it. When we arrived at the van, we divided what provisions we thought we could realistically use among us. It would require two of us to carry the tent, and we decided to take turns doing this.

By the time we reached the place where Paul's pack was lying we needed to rest. We put the tent down and took our packs off and sat down before gathering the contents of his pack.

After resting for about ten minutes, Brian walked over to Paul's pack to see how much work it was going to be to collect his items. He started laughing and told Milo and I to come over and see what Paul had packed.

When I saw what was scattered all over the side of the slope of the mountain, I could not believe my eyes. Paul's backpack contained only three items: Several pairs of long winter underwear, plastic ship models, and baseball cards.

In Search of the Unknown

I did not know just what to think at first. I just stood there looking at all of this wondering what he was thinking to pack such things for this kind of trip. Then Milo broke the silence, saying of Paul, "What a dumb ass!" Milo said we should just leave everything where it was, but I said we should get his pack and what underwear we could that was not too much work to reach.

The three of us decided since the contents had scattered over a wide area and it would be very difficult and time consuming to gather, that we would leave the models and cards.

I said that if Paul wanted the rest of the items, he could gather them on the trip out in a couple days. We collected the items we had decided upon, then put our own packs on and resumed the trip to camp. As we hiked the third person who was not carrying the old heavy canvas cabin tent carried Paul's pack.

When we entered the tree line where camp was established and a bonfire was now burning the others greeted us. We immediately told Kirk and Scott what had been in Paul's backpack and they laughed so hard they didn't think they would stop. When asked why he had packed those items, Paul once again only smiled and shrugged his shoulders. (Many years later, Paul would say that he had not packed the ship models and baseball cards. He claimed his younger brother had switched the contents of his pack as a joke.)

We completed setting up camp and took a look around the immediate area to become familiar with our surroundings before nightfall set in.

We learned many valuable lessons after the Clark ranch incident, one of which was to know the terrain around camp and any avenues of approach. Knowing where creatures might be coming from would allow us to be more aware of what to watch for and even to set up early warning techniques.

I had become very familiar with the story of the 1924 attack that Fred Beck had told. All of us knew the details. Looking at the upper end of Ape Canyon we could imagine the Sasquatch falling into it when Beck shot one of the creatures. We did not think about how isolated we were there should a similar situation occur with us.

Instead, we just behaved as typical eighteen-year-old boys and went about making dinner and joking with each other. That first night the joke was on Brian.

After our dinner was cooked, I took my plate and sat on a boulder near the tent. Scott came over and sat on a makeshift bench the college students had made.
This consisted of a thick piece of wood tied with rope to two trees.

Scott sat down, but instead of relaxing and eating his meal, he looked at me and got up and left without saying a word. I thought he had forgotten something, and went to retrieve the item.
While he was away, Brian came over and took the spot where Scott had intended to sit. Brian had just begun to eat his dinner when suddenly he fell backward with his feet going straight upward and his plate becoming airborne.

Standing right behind where Brian lay was Scott, with a hatchet in his hand and a big smile on his face. Scott began laughing hysterically. Brian got up knowing he had been the victim of a prank.

Scott had cut the rope holding the bench in place. Once it was released, Brian went flying. We all started laughing so hard we nearly dropped our own dinners. Milo looked at Scott and said, "Scott! You're one of us now!" Up to this moment, Scott hardly took part in the pranking that went on, and was usually the one it was directed at.

Scott said that he had finally had enough crap from Brian and decided to take the opportunity to get him when the chance came. We all congratulated Scott, telling him that this joke was worthy of anything the rest of us would do.

During the daylight hours, we conducted searches of the area. It was very rocky and too hard for any sort of animal sign, so we did not see anything interesting.

The final day of the Mt. St. Helens expedition, we woke to snow falling heavily. The fire was only glowing embers, and we got up quickly to gather wood since we would want to be dry for the hike to the van.

I went with Scott, Brian and Kirk to gather wood, and had quite a shock when we returned a short time later with arms full of branches. There, next to the fire pit, were Paul and Milo. Paul was pouring black powder on the hot coals while Milo was kneeling down close and blowing on the embers.

The four of us looked at each other briefly then began walking backward almost simultaneously without saying a word. We knew what was about to happen.

We had taken only a few steps when the inevitable happened, the gunpowder exploded with a big WHOOSH! Apparently Paul had a couple pounds of black powder in the pockets of his trench coat; he shot at black powder rifle competitions and had a lot of it.

Why either Paul or Milo thought it was okay to pour black powder on very hot embers was a mystery. When it exploded, a huge cloud of ash obscured our view of the two of them, but we could hear Paul coughing.

Evidently Paul still had not learned to hold his breath a few moments until the air cleared. Obviously, the lesson from the tire smoke a few nights previously had made no impression on him.

When the ash cloud cleared, both Paul and Milo remained in the positions they had been at the moment of the explosion. Milo looked dazed and we could see that both his eyebrows and hair on his chin had been singed off from the heat of the explosion.

Paul was coughing but otherwise untouched aside from being covered with gray ash. Soon we realized that both of them were mostly unharmed by the blast, except for Milo's facial hair. The situation became hysterical to the rest of us. The four of us literally fell down laughing at the two of them and the thought of what they had just done. We all still have a good laugh over that story to this day.

The snow was falling much harder now, so we decided that goofing around was out of the question. We had to get camp packed up and get out of the area.

We knew that if the snow continued at the rate it was falling, we could have trouble reaching Windy Pass and getting down below the snow level.

Paul, Kirk, and Scott were to carry the tent out since Brian, Milo and I had carried it up the slope of the mountain. The rest of us would blaze a path through the snow toward the pass.

In Search of the Unknown

As we travelled, Brian had chosen to stay with Kirk and Paul, and Scott caught up with Milo and me. Visibility had become difficult with such large snowflakes, but we still made good progress.

We started moving much farther ahead of the group carrying the tent, as they were moving slowly. The three of us made it to Windy Pass about twenty minutes ahead of the others and descended the slope much quicker than it had taken us to climb.

When we approached the place Paul had dropped his backpack on the trip in, the snow had turned to light rain. Scott wanted to see for himself the contents of the pack.

Scott just shook his head when he witnessed for himself the scattered contents of Paul's pack. He could not believe that Paul would bring such things on a trip like this.

We pushed on down the mountain to the van. Once we arrived at the vehicle, we looked and saw the rest of the group just beginning to descend the slope. I told Scott and Milo I was going to the restroom provided for hikers, and would be right back.

When I came out of the building, I heard Scott loudly saying, "Oh no!" I walked over to them, and asked what was wrong.

Scott pointed to the windshield of the van, and said, "That's what's wrong." Being bored the few moments I was away from them, Scott and Milo decided to shoot their wrist rockets and the home made steel projectiles we had made into the air. One of the projectiles returned to earth, landing in the center of the van's windshield.

The windshield did not break, but instead made three round cracks making the glass resemble a bull's eye. Scott said his dad was going to kill him, as this was a brand new vehicle. Scott's father was very nice, and I could never think he would get angry enough to scare Scott. Just the same though, I was glad it wasn't me!

Scott also discovered now that he had lost the gas cap, probably when we had stopped to fill the tank on our way in. I told him we could stop where we had gotten the fuel, to see if the cap might still be there.

This made him feel a little better, but not much. The white gas we had brought for the lantern also spilled inside the back of the van. We were very glad one of our other friends who smoked had not made the trip with us.

The thought of a smoker in the van with spilled gas gave Scott momentary visions of his father's new van going up in flames, and added to his anxiety.

When Paul, Kirk and Brian reached the van, Brian told Scott this series of misfortunes was his karma for the prank he played on him. Scott was not amused by this, and worried the entire trip home that day.

We stopped at the gas station, but the gas cap was nowhere to be found. Scott knew he was in big trouble when he got home.

As we reached town later that day, Scott dropped everyone at their respective homes. I was the last one to be dropped off, as Scott and I lived relatively close to each other. I assured him that his dad would not be too mad, and that things would be okay.

The next day at school I asked Scott how his father took the news about the windshield and gas cap. He told me that his dad only said that it was covered under insurance, and that it was no big deal. Scott said he was shocked by the lenient response and that his dad did not say anything more about it.

A few weeks later, Scott gave me copies of the pictures of the Sasquatch footprints we had found. I mailed them to John Green in Canada. Green wrote back a few days later saying the tracks looked pretty good. He asked if the picture had been taken using a flash.

Green never said anything else about our trip to Mt. St. Helens, except that it had always been a good place to search for Sasquatches.

That was our last outing during high school, and the last of the Pierce County Sasquatch Investigation Team.

René Dahinden and Dennis gates came to the Puyallup area once more to search. They wanted to see if there had been anymore Sasquatch activity where the Puyallup screamer incidents had taken place.
They first came to my home to ask me if I would like to go along and help them. I went readily, hoping to hear anything new about Sasquatches. I told them about our recent trip to Mt. St. Helens and that we had found a long line of footprints the first night there. René asked me to send him copies of the photographs.

Both men hoped that they might be able to again corner one of the creatures as they had done the previous year. They had brought the largest caliber rifle I had ever seen to take one of the Sasquatches down.

(William Jevning senior year of high school, age 18)

The three of us stayed alert for two nights searching, but there were no signs of the creatures in the area. Dahinden decided it was time to return home, so they took me home and said there would be more hunts I could come along on.

Most of our group of friends had already enlisted in various branches of the military and would be leaving home right after graduation from high school in June. As graduation approached, we did not think about the creatures roaming the forests.

(William Jevning at a friend's wedding fall 1977)

WILKESON WASHINGTON 1980

I had been stationed in Germany my first two years in the Army. I returned to Ft. Lewis, Washington as a Sergeant in 1979. I was happy about the assignment since it was close to where I grew up and my parents wanted me to stay at home while I was assigned to that post.

One evening in 1980, our longtime family friend Charley Welcome called my parents' home. Charley lived off the grid most of his life and would have been right at home as a mountain man in the 1800s.

I had known Charley for many years. He only owned a telephone two times. The first being when he worked for the National Park Service at Mt. Rainier when he had a house in Carbonado. The second time was near the end of his life in the same town where he owned a small engine repair business. It was quite an ordeal for Charley to make a telephone call. If he made the effort to call anyone, it was important.

When he called, he asked my mother if I were home. She told him I was standing right there. He said he wanted to talk to me. Charley asked me if I could come right up to his house. I told him I would be there within the hour.

When I arrived at his home, Charley had quite a story to tell me. He said earlier that afternoon a supervisor from the regional Game Department office had paid him a visit. This official asked him and his girlfriend, Maryanne, if either of them had been hearing any strange animal cries in recent weeks.

(My "home away from home" from 1977 to 1979 at Coleman Barracks near Mannheim Germany, this was 3rd platoon A-Troop 3/8 Cavalry)

Charley said he immediately knew something odd was going on, but acted like he knew nothing. He was familiar with the things I had seen and heard and thought this could be similar, so he thought the game official might tell him.

They both said they had not heard anything out of the ordinary and demanded to know why he was asking such questions.

This particular official knew Charley personally and explained that he would tell him, but instructed him not to tell anyone. If Charley were to say anything, the official would say that Charley had fabricated the story. Charley agreed to keep it to himself.

(A young Cavalry Scout, William Jevning at tank gunnery at Grafenwoehr Germany 1978)

The officer proceeded to tell them of mushroom pickers that had found something strange that had upset them greatly. The people involved had been searching an area a few miles on the mountain above where Charley lived. The place they were was commonly known as the alder flats. A herd of elk lived there most of the year.

The mushroom pickers had found a grizzly sign that morning. Two elk had been literally torn apart. The game officer showed Charley the report that had been made by the local game officer. He particularly recalled one part that said that the elk had been, "Dismembered without the use of tools."
Then the official mentioned the word, "Bigfoot", and Charley continued to act as if he knew nothing about this. The officer told him that they had a heavily armed team up at the site searching further.

(William Jevning dressed in old style Cavalry uniform as a member of the squadron color guard while stationed at Ft. Lewis Washington, a squad leader with D-Troop 3/5 Cavalry "Air mobile" Recon platoon)

He then said to both of them that if they ever heard anything to be sure to get in touch right away, and Charley agreed to do so. The game official then left, but before he drove away, he told Charley that it would be a shame if the power company discovered the location of their big power drain in the local area, pointing to the contraption Charley made to draw electricity from the nearby lines. The power company had been unable to trace where the power was draining, and assumed some short in the line.

Charley would face criminal charges if he was caught draining the lines for his own use without paying. So the official threatened him with this if he told anyone about the discussion they had that afternoon.

(While in Germany just prior to becoming a Sergeant, I got to be a tank crew member for a time, this was A32, I was its driver in 1979 for tank gunnery qualification)

Charley said he would tell no one, and then the official drove away. Charley then drove to Wilkeson to call me.

I had my shotgun in my truck and went up to the flats to have a look for myself. I found the place where the Elk had been killed. The game officials had taken the remains with them, but the ground was still covered with blood.

I searched the immediate area and found some very good Sasquatch footprints. I took a few pictures and decided to come back with a couple friends to look further the following day.

(Sketch courtesy Jerry Bishop)

When we returned the next day, we found more tracks from the direction where the creatures had approached the elk herd. Once I had the film developed, I sent the pictures to John Green. I also had written the account of events as Charley had relayed them to me and included them with the pictures to Green.

Green wrote back a couple weeks later saying the tracks looked pretty good, but he never said anything more.

I was very busy with my duties in the military in those days, and I did not have the time I would have liked to pursue my search for the Sasquatch.

RE-BIRTH OF THE PACIFIC COAST SASQUATCH INVESTIGATION TEAM (PCSIT)

By the mid-1980s I had gone through a divorce and left the military. A friend who lived near Vancouver Washington suggested I come and stay for a time to get away and clear my head. He suggested that if I liked it there I should stay and he would help me find a job and establish myself in the region.

I thought it was a great idea so went to Vancouver and started a new life. The Vancouver area and southwest Washington was not heavily populated in that time period and generally had a small town feel to it. It was a nice place to live.

About six months after relocating I met and became involved with a woman named Alice, who like me, was from the Puget Sound area of Washington. She had lived in another state for a number of years and had relocated to Vancouver near the same time I had.

We became a couple and spent a lot of time together. One evening while we were talking she remarked that it was nice that I spent so much time with her and her children, then asked if I had a hobby or something that I would like to do for myself.

The subject of the Sasquatch immediately came to mind. I cared for her and our relationship very much and didn't want to say anything to ruin what we had. I was not sure how she would take my saying I hunted Sasquatches! But I wanted to be open with her so I explained the involvement I had in the subject previously. She was enthusiastic about it to my great relief, and said I should start looking in our region.

(Our home in Vancouver Washington)

I was excited to begin working the field in this area encompassing Mt. St. Helens, as there was a lot of Bigfoot history in this part of the state. My first step was to write letters to John Green and René Dahinden telling them where I had moved to and to ask their advice about this region.

Both men responded very enthusiastically, Dahinden especially and said he was glad I was there adding that there had not been a Sasquatch hunter working that part of the state since Roger Patterson in the 1960s.

I felt like a kid in a candy store! I had the entire region with the richest Bigfoot activity history all to myself, and the pioneers of the topic supporting me. But where to start?

After discussing this with Alice, I decided to see if I could find out where the most recent sightings of the creatures had taken place. To do this I wrote a flyer asking about recent sightings of the creatures in the area. A co-worker made a stack of copies at his father's business. (Not too many copy machines in those days).

Alice's son Joshua and his friend David took the flyers around town on their bicycles and placed them under the windshield wipers of parked cars. Then I waited patiently for responses.
I did not receive any responses from witnesses. However, what did happen would set me on the path I was looking for.

I received a note from the editor from the *Post-Record* newspaper of Camas, Washington, asking me if I would consider being interviewed for the newspaper. I telephoned the editor and agreed to his conducting an interview with me and we
set a date and time. I had never been interviewed for anything for then public before so was a bit apprehensive, but interested so see what the experience would be like.

On the day of the interview, Kevin, the editor arrived as agreed. He was a very nice man and we had a great talk. I have never been a collector of Bigfoot paraphernalia, and I believe he expected our home to contain everything one could collect on the
topic. I had John Green's three books, René Dahinden's book, John Napier and Ivan Sanderson's books for reference and that was the extent of it. I think this caught him off guard, and he asked more questions than I think he was prepared to prior to the interview.

I asked him how he learned about me and my work on Bigfoot. He told me that one day he was walking into the courthouse in Vancouver when a police officer stopped him with one of my flyers in hand. The officer had found it under the windshield wiper of his cruiser and was about to throw it away, when he saw the editor and asked if he might be interested in it. Kevin took the flyer and contacted me regarding an interview.

At the conclusion of our talk, Kevin was very interested in maintaining contact with me. He told me he had enjoyed our discussion and said he would like to help me in any way he could with my work, adding that anytime I wanted a newspaper article published I only needed to call him.

The Camas newspaper published a three page article based on our interview, and this article was the catalyst that exposed my work in the region publicly and brought many witnesses into contact with me.

One of the people that subsequently reached out to me was a man named Carlo Sposito of Portland, Oregon. Carlo was primarily interested and engaged in the topic of UFOs, and cattle mutilations in particular. Carlo had his own encounter with a Sasquatch in the 1970s near Mt. Rainier and wanted to get in touch with a real, "Sasquatch Hunter". Back in those days we did not refer to ourselves as "researchers" or "investigators" we were simply Sasquatch hunters.

Carlo was a very intelligent man, and while he never said so, I believe he was highly educated. He became a very good friend and knew many people in the region who had told him of their own encounters with the creatures. He was interested in accompanying me during my field work when he was available.

Prior to meeting Carlo Sposito, the photographer of the Camas *Post-Record* newspaper had grown up in the town of Washougal, Washington which is next to the town of Camas where the Washougal river joins the Columbia River. The editor, Kevin, told me that his photographer knew many local witnesses of Bigfoot encounters and could show me locations where the creatures had been seen.

I began my searches near the locations where the photographer had shown me, and it was not long before I began finding evidence of the presence of Sasquatches.

Many of the encounters had taken place along the Washougal River, so recalling René Dahinden telling me too often closely watch waterways, I knew to search carefully along this river since the majority of sightings had taken place there.

The snow level was still low that time of year, and Alice and the kids would go with me to search along the Washougal River. We had some good outings to do field work and play in the snow.

On one of these excursions we found Sasquatch footprints leading up the hill in elevation. I had begun counting the prints and when I reached about 60 tracks, I heard what sounded like someone attempting to break into our vehicle, it sounded like someone using a tool to open a door, so we ran back to the vehicle.

When we came into view of the car we saw a brown pick up, one man standing near the car and another behind the wheel of the truck. I did not slow my run and confronted the man standing near our car and started a tirade, wanting to know what he thought he was doing! I am normally a pretty easy going person and have often been described by employers as very level headed, but having been a Sergeant in the Army my instincts kicked in and thinking this was an attempt to steal our car or some other nefarious behavior I was quite upset.

In mid-stride of my dressing down this man, who was dressed in regular civilian attire, I glanced at the man behind the wheel of the truck and noticed, he was wearing a uniform with a badge on his shirt. I now knew they were probably some local officials but kept right on with my questioning, not breaking stride and when I finished with the first man, went over and started in on the man in uniform! Looking back on the incident it is humorous now, but I was pretty angry at the time. The uniformed man apologized for the confusion and explained that this area was one they watched because a lot of stolen vehicles were brought there.

I asked him about the noise was we heard, saying it sounded like metal on metal and we thought someone was breaking into our vehicle. He could not account for the noise; most likely it was his PA system on the truck. He introduced himself and the man with him as regional game officials, and he knew who I was from the Camas *Post-Record* newspaper article. He was very friendly and asked me if I was conducting a search of the area, I said that I was. He asked if I was having any luck, and I did not mention the footprints we had just found up the hill, answering only that we had just begun but were hopeful.

He expressed interest in what I was doing and gave me his card saying that he would be happy to watch my vehicle anytime I was doing field work, I just had to call him and tell him where I was and he would make sure no one bothered my vehicle. I saw the game officials from time to time during my years of field work in the region; we always had a very cordial relationship although I never told them or anyone where or when I was doing my field investigations.

After talking with the game officials, we returned to the line of footprints and attempted to follow them. However, the snow was getting deep and we were not able to follow them any longer so decided to return home.

We made other trips to the same area and found a line of Sasquatch footprints on another occasion. The tracks measured 18 inches in length so I knew we had the same individual in the area.

In March of 1988 we made another trip to the same region along the Washougal River. I was hoping we would find more tracks of this Sasquatch while there was still snow on the ground. The area was not conducive to footprint impressions, it was very rocky and the road was hard packed gravel, so I wanted to take advantage of the snow as a medium for footprints.

While we drove along the river, I noticed how beautiful the water was. It flows on bedrock and there was no silt and the water was clear and blue green in color and reminded me of the Ohapenacosh River where I fished as a boy.

Suddenly I noticed movement out of the corner of my eye, and saw a massive gray/white Sasquatch next to the water. It took a couple of quick steps disappearing into the thick foliage. Everyone with me except Alice saw it, and I immediately tried to move to see if I could still see the creature but it was gone. The river at that location was deep and fast moving enough as to make it impassable for crossing on foot and there was no other way to get to the spot where the creature was.

I asked each person to sketch what they saw without discussing anything; each witness drew exactly the same thing. The creature had its left side to our position and its back as it quickly moved away, it reached out and grasped an alder tree with one hand and then disappeared into the brush.

I made notes about what happened and we drove on hoping to find tracks of the creature in the snow, but the snow was completely gone so there were no visible signs of the Sasquatch.

I drove the following day to the Camas *Post-Record* newspaper, and told Kevin the editor what we had experiences the previous day. He said he would publish an article about our experience. I was hoping the article would entice possible witnesses in the area to contact me so I could gather more information and determine if this Sasquatch was alone or there were also others in the area.

March 10, 1988

Bigfoot plays peek-a-boo with team of searchers

A team of searchers say they spotted at least one elusive gray-haired Sasquatch near Dugan Creek in the Upper Washougal River area in late February.

William Jevning, head of the Pacific Coastal Sasquatch Investigation Team, said Monday that at least three members of his team saw a large gray-haired creature across the Washougal River near Dugan Falls Feb. 28 during one of the group's ventures into the rural area northeast of Washougal.

Jevning also said the group saw tracks that indicated there may be at least one or two other Sasquatch living or traveling in the area.

"We've been kind of keeping an eye on that area for a long time," Jevning said. "Then, when we were driving up there to look over the weekend, across the river we saw some movement and there was this enormous thing trying to get out of our line of sight."

Jevning said the creature must have been between 9 and 11 feet tall and weighed an estimated 1,500 pounds.

Members of the search team photographed more than 60 tracks that were about 20 inches long and several inches deep in the hard ground, Jevning said. They also found tracks of a smaller creature, believed to be another Sasquatch living in the area, he said.

Jevning, a former Army reconnaisance specialist from Graham, Wash., has spent the past two months

• Besides his hide-and-seek games near Washougal, Bigfoot is now the star attraction at Knott's Berry Farm in Southern California. The story is on page B4.

searching for the legendary Sasquatch — or Bigfoot, as it is known — in the Upper Washougal River area and into the Gifford Pinchot National Forest.

Jevning restarted the Pacific Coastal Sasquatch Investigation Team after a 10-year hiatus during which the team, which originally began with his friends in Pierce County, was disbanded while Jevning was in the Army.

The team has been collecting information about Sasquatch sightings and tracks in southwest Washington for several months. Jevning said so far several new people from the area had joined the team.

Information about Sasquatch sightings or tracks should be sent to Jevning, PCSIT, Suite 298, 1701 Broadway, Vancouver 98663.

During the Feb. 28 expedition, Jevning said about five people in the team were driving cars on the unpaved portion of Highway 140 past the Dugan Falls area when three of them — including Jevning — noticed something large moving slowly along the river bank. The creature apparently was trying to get away from the humans, Jevning said.

Alice and I took the kids and drove to a higher location where there was still snow above the area where we had the sighting hoping to find evidence. While Alice and the kids were playing in the snow, I saw another couple with their children nearby doing the same thing as us. I had my 35mm Pentax camera hanging around my neck at the ready. The man from the other family walked over to me and introduced himself, asking me if I was getting any good pictures. I responded, "I hope so", and apparently the way I responded prompted him to ask me what I was interested in taking pictures of.

I explained what I was doing and about the creature we had seen recently. An odd look came over his face and in a very serious manner told me that he had once seen the very same creature. His encounter had happened 17 years earlier while he and a group of friends were in their hunting camp one evening. The men had been standing around the camp fire discussing the next day's hunt, when this massive gray/white creature emerged from the darkness of the night and stood watching them at the edge of the fire light.

Neither the men nor the creature moved for a time, and then the creature turned and disappeared once again into the darkness of the night. He was visibly shaken even this many years later. His wife had been approaching our location as he was telling me his story, and when she joined us told her husband that he had never told her that story. His response was that he thought she would have thought he was crazy. She told him maybe, but she still would have listened to him without judging him seeing how it affected him even this many years after the event.

As I came to interview more and more witnesses in the region, I noticed how the Sasquatch depiction in John Green's books was not entirely accurate. Up to this time the creatures had been generally described as loners, primarily vegetarians that were shy and wanted solitude. This picture began to drastically change as I gathered more information through witness testimony and my own observations.

I would periodically make visits to John Green and René Dahinden's homes or usually through letters, tell them each of my findings. Their responses were typically that they did not know enough about the creatures to make solid determinations and had simply made assumptions based on witness testimony.

I determined that this was partial information at best and thought a lot about the physical needs of creatures of the size of the Sasquatch and how they would acquire what they needed to survive in the environments of the Pacific Northwest.

I also learned that there were far more witnesses that never spoke about their encounters with the creatures than those who did. My estimate based on years talking with many people during the 1980s and 1990s was that there were likely ten or more witnesses having experienced an encounter with one or more Sasquatches and never telling anyone for every one witness that had told someone.

I knew there was a lot more information out there we did not know that may help provide a clearer picture about the creatures and why it was so difficult to get close to them, let alone prove they were real. I will not go into detail on this topic in this volume, as I am going to discuss it in some detail in another book project.

After having the encounter along the Washougal River in March of 1988, I was contacted by a representative from Knott's Berry Farm who was in Portland searching for someone knowledgeable about the Sasquatch. I was working a graveyard shift at the time and when I arrived home from work one morning, Alice had told me that this person from Knott's Berry Farm had obtained our phone number from the editor of the Camas newspaper, and she wanted to talk with me. I called the representative and she told me Knott's Berry Farm was in the process of making a new attraction and wanted someone that was knowledgeable about Bigfoot, and asked if we could meet. We set a date and time for her to come to our home.

She had planned to meet with me one morning shortly after I arrived home from work so I would not have to stay up long and could get to bed. When I met her initially the first thing I said was that if Knott's Berry Farm was making anything but a serious portrayal of the Sasquatch, then she could leave and they would have to find someone else!

She assured me that they wanted nothing but a serious portrayal of the creatures. They had financial backing from Alaska Airlines in the sum of ten million dollars for what they called "Bigfoot Rapids", which at the time was to be the largest man-made white water ride themed attraction In the center of the "ride" was vegetation from the Pacific Northwest and a ranger station in which visitors could come see displays of all the animals of the Northwest including the Sasquatch. They wished to hire me as the technical consultant regarding the section on Bigfoot. I agreed and she told me she had to get back to the park and would be in touch with the details.

Hal Stoelzle/The Register
Will Jevning holds a cast of a Bigfoot 'footprint.' An exhibit of the alleged creature opens at Knott's Berry Farm on May 27.

Bigfoot believer overseeing upcoming Knott's exhibit

A couple of days later I received a call from the representative I had met with. She said they wanted any footprint casts and asked if I could get copies of newspaper articles of witness sightings, plus any photographs, maps of sighting locations and so on. I said I could provide what she had asked for. She then asked what my fee would be. I had absolutely no idea what to respond with, as far as I knew up to that time no one had ever been paid to do anything Bigfoot-wise. I told her I didn't really know, and had not thought about a fee amount or even begun thinking I would be compensated in any way for helping with the project. She asked, "How does three thousand dollars sound?" I said, "Sure that would be fine." Thinking back I could have probably asked many times that amount and received it, but I was just happy to be asked to help let alone get paid for doing it.

I called John Green and René Dahinden explaining what I had been asked to do, and asked them if they had any materials I could purchase from them for the project. (This was before I had begun casting Sasquatch footprints so I did not have any casts.) Green told me to call Grover Krantz who had copies of a number of tracks he was selling. I phoned Krantz and he said he had about half a dozen tracks he would send me, so that part was completed.

I made plans to drive to both Green and Dahinden's homes. From Green I purchased some of his books, and Dahinden sold me copies of more than 600 newspaper articles. Dennis Gates had previously owned and operated a clipping service. When he quit running the business had sold everything to Dahinden, so for $20 he sold me all those articles. I then purchased three dimensional maps from a local map maker in Vancouver and placed markers to show where Bigfoot sightings had taken place in the upper north-western United States.

René also gave me a number of postcards he had made from the Patterson film, along with some of my own field findings and photographs. Alice and I were flown to southern California as guests of Knott's Berry Farm.

We spent a couple days at Knott's Berry Farm. I had shipped the display materials before we went there so they would have them for the project. My main job was to help promote the attraction by doing news conferences. We all see news conferences on television; a big room full of reporters asks questions of someone. This way of conducting news conferences is reserved for important people like the President and such. Everyone else gets something quite different!

I spent two grueling days talking in a small windowless room with reporters one at a time! I do not even know how many I spoke with. Most were very interested in the subject of Bigfoot, not knowing much about it. A couple seemed to have a chip on their shoulders toward the subject, but even they came around—if not completely convinced in the creature's existence—at least enough to write a good piece on me and the project.

We got to spend a day enjoying the park and all the rides and attractions during employee day courtesy of park management. Overall it was a great experience except that Alice who got fluid on the knee and had to be in a wheel chair. I wish that it would have been a fun experience for her too. She was a real trooper throughout the experience.

In 1987 as we had discussed my starting work in the southwest region of Washington State, I wanted to create a network to help in the collection of information. Some of our friends had expressed interest in becoming involved, so immediately I thought of the semi-organized effort we had made among my high school friends after meeting John Green and René Dahinden in 1975.

I decided to re-create the old PCSIT, now calling it the Pacific Coast Sasquatch Investigation Team. We had a friend who worked out of the governor's office as an advocate for handicapped children, and was knowledgeable about creating an organization. We decided then to make it a non-profit organization to demonstrate that we were not out to make money for our efforts, and this is what the new PCSIT became registered as. For many years afterward the PCSIT remained the only organized effort dedicated to resolving the question of the existence of Sasquatch its kind.

It was not long before the PCSIT grew requiring a governing body. In the creation of the organization, and being registered as non-profit, we had created a constitution with all the necessary rules covering management and operations. We established a board of directors who were voted into office by the members, which numbered nearly 100 after we became established. We were very fortunate to have attracted very professional and talented people especially among the board members.

As members joined our ranks I established field teams and widened the scope of my own investigations. South of Mt. St. Helens is a huge area covering 3000 square miles to the Columbia River, and it was a daunting task just learning the region let alone finding Sasquatches.

Often times we would get leads on recent encounters and have to spend a lot of time navigating vast areas of forest in systematic searches, most of the time coming up empty handed. Anyone who thinks becoming involved in this topic is exciting does not realize how much of it results in disappointment and sheer boredom.

(Board of directors PCSIT 1990)

It has always been a very difficult task conducting quality field work, especially having a 40 plus hour work week and all the other mundane particulars of day to day life.
Most often this leaves far too little time to conduct field work properly. I kept strict log books that chronicled the amount of time, exact locations, miles driven; fuel used, and detailed results of my searches. During the time from 1988 to 1998 I documented more than fifteen thousand hours of active field work in Southwest Washington State.

That was a tough decade of work utilizing all my spare time. It took its toll, but I also gained very valuable information over those years.

(Rene' Dahinden, William Jevning and Carlo Sposito Richmond British Columbia mid-1980's)

1989 was both a big year for my field work, and a year of change. One morning in early June, I went out to the front of our home to get the newspaper and sat down with it on the front steps. I looked at the folded papers headlines, and then turned it over to see an article about a family in the nearby community of Yacolt that had reported an encounter with a Sasquatch just three days prior. I won't go into detail about the events that followed as all of that is covered in my book *Haunted Valley*, but that day began a nine month investigation by me and the PCSIT.

A month later Alice and I had broken up and I moved temporarily to Ridgefield. It was a very difficult time for me, and to fill the void being apart from Alice I spent a lot of time on the new investigation that rapidly developed into almost daily encounters and new evidence being found near Yacolt.

One of my goals was to establish local contacts throughout Skamania and Clark Counties. Communications were not nearly as fast as they are today with computer technology and cell phones. To get reports to me as quickly as possible Dahinden had told me it was necessary to have locals who would learn of encounters and in turn alert me to them. Then I could minimize time in getting to the scene of an encounter and collecting evidence before it was destroyed or degraded to uselessness.

My friend Carlo Sposito told me one day he knew someone in the town of Carson, Washington, some distance east of Vancouver and in eastern Skamania County who was the local Bigfoot enthusiast. His name was Datus Perry, and locals that had encounters with Sasquatches would nearly always call Datus when they saw something. Carlo had spoken with Datus and told me Datus wanted to meet me.

I was excited at the prospect of having a contact in eastern Skamania County as it would greatly reduce my time in getting sightings there. Carlo and Datus had arranged for Carlo and me to go to Datus' home.

During our drive to Carson to meet Datus, Carlo briefed me on what he knew of Datus, saying he was a bit "eccentric". As we turned the final corner leading to the home of Datus Perry, we saw next to his mailbox a nine foot hair covered plywood replica of what he later told us was the Sasquatch he saw in Alaska.

Carlo told me that Datus was very excited to meet me as I was a "real Sasquatch hunter", and placing his replica out was his way of greeting me. His replica was covered in bear hair; its head was very pointed with arms and hands barely visible as they were flat against the sides. This gave the appearance of a burnt tree stump, which later after listening to the story Datus told of his "sighting", I became convinced was exactly what he had seen—a burned tree stump.

(Datus Parry foreground wearing white hat and beard)

We parked in the driveway of Datus and his ex-wife Lillian's house. The house and surrounding yard were something of a free range area for chickens and goats, and assorted junk cars and other odds and ends of what appeared to be long forgotten projects. Datus came out to greet us. He was old, Carlo had told me, but even Datus did not know exactly how old.

Datus was very thin, had long white hair and a beard with bright quick eyes. He was somewhere between 79 and 85 years of age. (He changed his age each times he was asked.) He claimed to be a relative of Admiral Perry, and said he had lived most of his life in Carson except when he traveled to Alaska.

Carlo asked him to tell us the story behind his "Bigfoot" replica and he seemed excited that we would take an interest in his story. He told us that as a young man in Alaska, he was hunting and trapping on a river he didn't recall the name of. He was in his canoe one day, and spotted the creature a few yards from the bank of the river watching him. Datus said it looked just like his replica; it just stood there watching him, never moving. He said he stopped and it watched him, motionless for about twenty minutes. He then decided to move on, and would see if it was still there when he returned.

When he got to the part of his story where he, "Paddled his canoe up a waterfall", I lost interest. I thought up to this point his account was interesting, although his replica did not look much like a Sasquatch, I realized that each person's perceptions are slightly different. However going up a waterfall in a canoe was just too much for me to accept as reality, and I switched the course of our conversation.

I asked him about any recent Sasquatch encounters he was aware of in this area. He said he would be happy to tell me and asked us to come into his house. Inside, we met Lillian who was very nice but did not participate in our conversations. Datus showed his pride and joy—a flip chart that he said was the absolute proof that science was waiting for regarding the existence of Sasquatch. This "absolute proof" was simply a collection of drawings he had made, some photo copied pages from John Green's books and a few of Datus' ideas which did not amount to much.

I held my tongue through all of his presentation of his "irrefutable proof". I felt that his acting as a collection point for sighting reports and witnesses outweighed this silliness. Datus was adamant that since he claimed to have seen a Sasquatch that science would have to accept that they existed. This was his irrefutable proof. I did not make any comments, but stayed polite, Carlo asked questions which seemed to please Datus. Datus later called Carlo to ask why I did not comment on his "proofs". Carlo simply explained that I was the kind of researcher who thought about things for a long time before commenting, and this explanation was enough for Datus who asked when we were coming to his home next.

Through our first visit with Datus, I learned that he was a self-promoter, who would tell literally anyone who would listen for more than a brief moment to what he had to say regarding the Sasquatch and more so of his own "encounters". But again, I felt at that time he could be a good resource for our work—if people told him when encounters occurred then we could get onsite quickly. My thinking was that through Datus we would get to new sighting locations, and then take over investigations. Datus never seemed to do anything with information local residents told him about. It appeared to be an ideal situation even if Datus was a bit "eccentric".

We began to learn soon enough though, that the information Datus had was of little use. Carlo would usually call me saying that Datus had phoned telling him that someone had seen a Sasquatch, and then we would drive there to investigate. Once we arrived at Datus' house, he would insist on driving us to the location. Datus had two working vehicles among many that actually were inoperable. One of the vehicles was an old junk-filled panel van, with only two seats and filled with junk. The other vehicle was an old Rambler station wagon that seemed to be composed of more rust and mold than steel.

Both vehicles were filled with various kinds of junk car parts and just plain trash. Riding with Datus Perry was an experience not embarked upon lightly! He was not the typical careful older driver one would have expected. Riding with Datus Parry was a white knuckle experience most would never wish to repeat. He normally drove at break neck speeds that we didn't think his vehicles were capable of, the whole while mumbling to himself.

When we would arrive in the vicinity of where the sighting in question was supposed to have happened, we would then learn that either the event had taken place years ago, or that Datus did not know where it had occurred. In almost every instance of an alleged sighting of a Sasquatch Datus took us to, there was nothing to find or see. Carlo and I did see a lot of the area in that part of the gorge neither of us were familiar with, and kept places in mind we wanted to look at without Datus at the wheel!

To compound information problems, Datus never wrote any information down. If someone told him of a sighting, he would often not recall more than the person's first name and never recorded phone numbers, addresses or any particulars of witness accounts. I began to wonder if he really was getting legitimate information, or if he was simply making things up to keep Carlo and myself coming to see him.

I was considering cutting off contact with Datus, when I spoke with a person at the local grocery store who claimed to have encountered a Sasquatch and said they had indeed called Datus with the account. So I decided to continue my association with him in the event good information came from his position in that community.

One day I did come into one of the best witness accounts of my years as a Sasquatch investigator, but not through Datus Perry. Carlo had been called Datus. He said he had been contacted by a witness of an encounter in the vicinity of the town of Stevenson, Washington. Carlo called me and we decided to go and talk with the witness, but as usual Datus had no contact information and was not even sure who the witness was. He took us to an old logging road that had not been used for any vehicular traffic for many years. I felt that if nothing else it might be a good area to explore. We drove to the place then walked the rest of the way on the road searching for signs of Sasquatch passage; Datus could walk a long way for someone his age. He seemed to wish to be in this location for some other reason than the supposed encounter that got Carlo and I there in the first place, as he kept talking about lava tubes and his exploration of them.

I became disgusted with the lack of purpose that day, and decided to look where I wanted to—even if it meant going somewhere other than where Datus was heading. Not long after thinking this, two young men riding dirt bikes came into view heading in our direction. They stopped when they reached us, and I decided to ask them if they had any knowledge of anyone seeing a Sasquatch in the area. One of the men, Kevin Gerde, told me that in fact he did. A friend of his had seen one of these creatures just a week prior, and he believed his friends story because he had known him a long time and considered him was very honest.

I talked to Kevin for a while and told him who I was and some of my experience, and he agreed to arrange for me and his friend, Hugh Brown, to meet the following day. Carlo wanted to meet Hugh with me so we decided to interview Hugh when Kevin called me and gave me the specifics. Datus did not wish to go with us and stayed home. Carlo and I drove to Stevenson the following day and met Hugh Brown. Hugh told us the best way for him to tell us the account was to go to the location where he and his friend, Jeff Strough, had seen the Sasquatch. I recorded Hugh's story while we drove to the location, which was just across the Columbia River at an area called Wyeth near the Bridge of the Gods in Oregon. Hugh's story was fascinating. He said that earlier the previous week Kevin had told him he should go to Wyeth to look at some steam vents he had seen there. Wyeth was a small community long gone by then. It had been a logging community but no buildings were left. The "steam vents" were actually vegetation on an old logging landing which had been buried, and the heat caused from the decaying vegetation would turn water from rain to steam. The steam then would escape from a small opening in the ground. Kevin had mistakenly thought he was observing volcanic steam vents.

Hugh and Jeff drove to the location to look at the vents. They parked and walked to the spot. After looking at what Kevin described a few minutes, they were about to leave when they heard a loud "roar" from the slope below them. They looked at each other confused; neither man had ever heard such a sound before. They watched and listened for a few minutes, then heard another roar but closer this time. At one point they saw an animal downhill from them in a small open area in the forest. They thought it was a bear but could only see the head and shoulders. The creature making the loud roar seemed to be coming toward them, as the sound kept getting louder and closer.

Jeff decided he didn't want to be standing there if a bear was coming toward them, and was afraid it would attack them. Jeff ran for the safety of their car. Hugh stood there trying to get another glimpse of the bear. Suddenly from the direction of the roaring, a deer sprang from the brush. The deer obviously terrified of something hardly even noticed Hugh and stopped close enough that Hugh said he could have reached out and touched it. Then the deer ran off into the forest, and something else came out near the place the deer had. Less than fifty feet from Hugh stood an enormous hairy man-like creature, Hugh had heard stories of Sasquatches and knew that this had to be one. The creature stopped momentarily, and then ran directly for Hugh. Hugh said he must have been in shock at what he was seeing, but could not move and just stood there.

Hugh thought that was it. He was going to be attacked by the creature. The Sasquatch got to within about twenty feet, still slightly below Hugh on the open slope and stopped. It stood there staring at Hugh, and Hugh stared back at it neither of them moving. Hugh said he estimated that this went on for about twenty seconds, before the creature simply turned to its right, and casually walked off into the forest.

Hugh got his wits about him by then. He ran for Jeff and the car and they got out of there. Hugh and Jeff drew what they saw as best they were able; I put the drawings in my book *Notes From the Field, Tracking North America's Sasquatch*. Datus had nothing to do with our getting this account, but we may not have come across it had we not been on that old logging road that day.

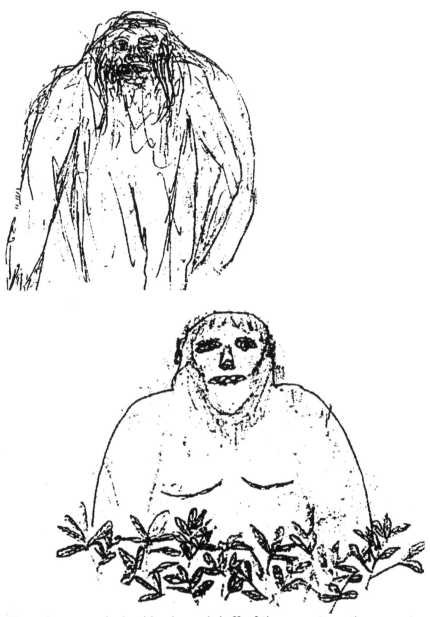

(Drawings made by Hugh and Jeff of the creature they saw)

Datus only gave me one piece of evidence over the four years we associated ourselves with him. It was actually something he did not consider important. I had occasionally asked him certain questions, to see if he had actually seen anything that was important evidence. I once asked him if he ever saw rocks stacked up anywhere. René told me to watch for these in any of the areas I searched to see if this was evidence we needed to research further. Datus said he knew of one place where there were rocks as I had described, but they were not done by Sasquatches. He thought perhaps native Indians were responsible. I told him I would like to see the place anyway, and for once he actually took us to a place I wanted to see. We again drove as far as possible, not far from Stevenson, Washington and proceeded the rest of the way on foot. Datus took Carlo and I as far as he could go, then told us how to get the rest of the way to the site. At first I was afraid he had sent us on yet another wild goose chase, and then we found the place. There were a total of nine large piles of rocks, and I took a number of photographs as they were similar to those René had shown me in Oregon.

Carlo also suspected the rock stacks may have been left by Indians, but I thought otherwise. We finished photographing the piles which were approximately five feet high and between six and ten feet across at the base. I later talked to some Indian friends in the area about the piles, and as I has suspected they told me Indians had not done this but it was the work of Sasquatches.

My thinking of Datus was a little better after this. It was a big find and I wrote Dahinden about it. This was the first find of rock stacking outside of Oregon. He was very interested and told me to fully document it so we could discuss it when he next came to visit me.

(My photograph of two of nine piles of rocks Datus took Carlo and myself to)

I thought this validated my thinking that some good might come from our association with Datus after all. However, things remained the same. He did not records details when he said someone talked to him about Sasquatch encounters. This time he called Carlo about a sighting by a helicopter logging crew. Supposedly, the pilot saw a Sasquatch sneaking up on the lunches of the loggers. According to Datus the pilot called the foreman who ran to where the lunches and other tools were, and scared the creature away. When Carlo and I got to Datus' house, as usual he had no names or any contact information. He took us to where the logging had taken place, but the site looked much older than Datus had described. Datus could not climb over all the tree limbs and debris, so we told him to stay with the car while we looked around.

(This is another of the rock piles Datus showed Carlo and myself)

I did find a weathered line of Sasquatch tracks, I measured them to be approximately seventeen inches long and six wide. The tracks were three to four inches deep in places where the soil was soft. The tracks would have been very good if we would have been there sooner. I took a number of pictures, and Carlo and I searched further hoping to find better tracks.

We did find a rock formation that concealed the entrance of a small cave. The cave roof was no higher than five feet, and it went back in approximately twenty feet. It was dry inside and Carlo and I could not stand so we went in on our hands and knees. There were fresh leaves and ferns on the bottom of one area, appearing to be bedding. I had seen Sasquatch bedding before, but it is something they rarely do and this could have been another instance.

The cave was recently used by some creature. We thought since there were Sasquatch tracks close by that one may have indeed been using this cave and bedding. I thought we should check the spot once in a while, and Carlo agreed. We decided not to bring Datus on follow up checks of this place, as he was no help to us. We did not find anything more there, and the vegetation that was fresh inside the small cave was never replenished. I told Carlo that the logging activity probably chased away the Sasquatch that may have been using this place. He agreed and we checked it a couple more times then stopped going there.

This was the first time Datus had actually taken us to the site of Sasquatch activity that we could verify, but the time frame was not as Datus had claimed. By this time, Datus had been taking Carlo and me into the wilds of Skamania County for a long time, with very little results and my patience was again growing short. Once again Datus called Carlo and said that Sasquatch footprints had been found near the Washougal River. I was excited about this because it was near where we had seen the big gray Sasquatch and footprints.

I told Carlo that this might be very good since my own experiences at that location were good, and we should go right away. We drove to Datus' home per his request, by now he would not take us to locations unless he drove himself. Carlo and I once again agreed since I felt this was worth the effort, all based on my personal experiences there. And Datus said he had the exact place marked on a map by the person who found the tracks.

(I discovered a long line of 18 inch Sasquatch footprints, unfortunately they were about a month old and in bad condition)

That day began with very heavy overcast clouds, and I knew it was going to rain and took rain clothing. I was hoping we could get to the place before the rain began, so we could cast some of the footprints. Datus insisted driving his old Rambler station wagon, which meant Carlo and I could breathe with some relief. When he took us in his van, Lillian went along. She would sit in the only other seat besides the drivers, leaving Carlo and I to be flung around the back of Datus' van along with all the junk he had in there. Those rides were no fun at all!

Instead of Datus driving highway 14 directly to Washougal and saving time, hopefully ahead of the impending storm, he drove all the logging roads through southern Skamania County. By the time we got half way along the journey there, a cloud burst happened. When we got to the place where the footprints were supposed to be, it was pouring so hard no tracks could have survived. I was now in a foul mood at Datus for costing us, our opportunity to cast the tracks. Now as usual, Datus did not know exactly where he was going. The map he had was marked by him and not the person who found the tracks. This made me angrier, because the whole day was being wasted.

Carlo sat up front with Datus, trying to help decide where the location was and attempting to glean any information he could from Datus. They finally agreed on the place to stop and look, and we got out and donned our rain gear. What finally made me decide never to go with Datus again was when he got into a furious argument with Carlo over a mud puddle! Datus insisted a mud puddle (full of water) was a Sasquatch track. The puddle was very round and clearly just a mud puddle—and even if it had been a track it had three inches of water in it. Carlo was very diplomatic, but after a while even he lost his temper and a loud argument ensued.

(Datus did show Carlo and myself around Skamania County, this is one of many lava tubes located throughout the region south of Mt. St. Helens)

I was wet, cold and fed up by this point. Rather than be involved in their argument I walked into the forest to look around. The area had very little under brush, and there was a short logging road there. I walked along this road that had a five foot high pile of gravel at its end. Right in the middle of this pile of gravel, were a very well defined left and right seventeen inch set of Sasquatch tracks! The rain had not destroyed them because they were in gravel, and I yelled for Carlo and Datus to come and look at the footprints. The Sasquatch that made them was very heavy; I stood on the gravel near them and made no impression at all.

Datus was in a huff and went back to his car. He said he was done for the day and was going home. Carlo came over to look, but he too was now very upset and even though we found tracks he wanted to go home and cool off. All that we went through and neither of them were interested once we found footprints! For me that was it, I decided to start going on my own again.

After this day I met with the PSCIT board members and we discussed Datus. We made him an advisor to the board of directors, mainly an honorary position as he was a nice old man, but no use as a local contact. We wished to remain friends with Datus and not offend him. He did call once in a while supposedly having a sighting location and sometimes we would check it out, but most often we ignored his information as it never resulted in anything.

In the fall of 1990 I was temporarily laid off from my job so had time to do other things. During this time I became acquainted with Jack Livingston. Jack was a single father of two very young boys who was between jobs at the time. We met through a mutual acquaintance and became friends quickly. We began discussing the subject of Bigfoot. Jack had many interests and Bigfoot was one of them. He began going with me on my trips to my field locations, and soon became very involved with my work.

The PCSIT board of directors soon voted in favor of the creation of the position of Director of Marketing, and named Jack the Director. His responsibilities were generally the promotion of the group and to assist in other functions such as fund raising.

(Jack Livingston and William Jevning Kingston Washington 1990)

I always made a habit of stopping and eating or buying things at shops in communities near where I was conducting my field work. I knew from the example Al Hodgson provided during the Bluff Creek years that local merchants were often the people who witnesses first told their encounter stories to. My hope was to establish an information network among local merchants to help me learn of anything fresh happening in those areas.

One merchant was Kevin Landacre who owned the Eagles Cliff store and campground just south of Mt. St. Helens, located a few miles north of the town of Cougar. I got to know Kevin and would make a regular habit of visiting him while in that part of the circuit I was working.

On one such visit in the spring of 1991, I stopped by prior to my leaving the forests nearby Eagles Cliff to chat with Kevin and see if he had heard of any recent Sasquatch sightings. It was just before sunset that day and I had taken a map out to show Kevin where I had been working. We were on the side of the building that housed his store, and thought we were having a private conversation.

We noticed that about sixty people from the campground had gathered around us, listening intently to our conversation about Sasquatch activity in the vicinity of the store and campground.

Kevin whispered to me that he was a bit nervous with all the people surrounding us, and asked if I would mind taking them to one of the nearby communal bonfires and talking with them. I said that I didn't mind but asked him why was he nervous. He owned and operated the place! He should be used to lots of people being around. He said usually he had the store counter between him and anyone coming into the store, and didn't like crowds.

He announced to the gathered group who I was and asked everyone to follow me to one of the fires, saying that I would be happy to talk with them and answer any questions they had.

This was a much unexpected event, and something I was not sure I was ready for. There was quite a mixture of people in the group assembled from many differing backgrounds as far away as England. I spent a couple hours chatting with them, all asking very good questions, and thoroughly enjoying the talk. Several were even videotaping the impromptu talk and I knew it must be getting late so I looked at my watch, and knew I had to get home. I thanked everyone for chatting with me and asking so many good questions.

In Search of the Unknown

During my long drive home that night, I thought a lot about how well the conversation had gone, and that not one person laughed or made fun in any way of the subject of Bigfoot, but had responded very enthusiastically.

When I brought that evenings events to the next board meeting of the PCSIT, Jack thought we might have a great opportunity at Eagles Cliff to promote our work and to tap into unreported accounts of witnesses of the creatures who might not otherwise know where to go if they had an experience to share.

The board of directors unanimously voted to draw up a plan for an event, providing Kevin the owner would allow us to stage something on his property. It was decided that Jack and I would visit Kevin and make the proposal for some kind of event to take place at his store and campground.

I was very apprehensive to ask Kevin about doing an event, not having any experience doing anything like it but Jack was confident and had an idea what we should do.
When we arrived at Eagles Cliff, I introduced jack as our Marketing Director and that we had a proposal to present. Being out of my element, I stepped away while Jack and Kevin discussed our idea.

When they had talked for some time, Kevin came to where I was looking at some items in his store and told me we had a great idea and that he was in 100 percent and excited about what we planned to do. I thanked him and said I looked forward to the event, and Jack and I then left to return to Vancouver.

Jack was very excited on the trip back to Vancouver, saying Kevin also was very excited about the prospect of attracting more business to his location, and promised to assist us with whatever we needed from him to make the event successful. Now we just had to come up with an idea of what to do!

We had discussions during many of the following months of PCSIT meetings, and finally decided to have a BBQ and potluck. The idea was to get those attending to become involved and feel like part of the event. We chose the following summer to have the event, as we wanted enough time to plan and make sure everything was in place for attendees to enjoy themselves.

My field work continued while the PCSIT planned the upcoming event, and I made an important discovery in July of 1991.

By that summer I had gathered a lot of movement data regarding the particular group of Sasquatches I had been following since 1987 around the watershed region of the Washougal River system. I watched one day an old documentary made in 1972 about the area and a group looking for the creatures in the area based on witness accounts. What struck me was the places they were looking in the film were the same locations I had been finding evidence.

At one point in the film, the group had attempted to follow the Washougal River into its head waters region, and had a very bad experience in the attempt, not being able to reach their destination. This intrigued me as that particular area was on my list of places to search as it was central among locations I had received reported witness sightings and where I had found Sasquatch footprints on numerous occasions.

In Search of the Unknown

Jack went with me on at least five separate attempts to enter the headwaters area, it was one of the most difficult places to get into I have ever experienced. In what seemed like the obvious way to travel into the area, going right upstream we encountered sheer rock walls on either side of the channel. These were more than one hundred feet of vertical rock, totally impassable. So I began looking for ways to bypass this obstacle. Eventually after a number of separate trips to this place, we found a way into it. We had to approach the area from an entirely different vantage, climbing to the top of a ridge system that pointed toward the watershed of the river. The best (and only viable) access was to hike along the top of the ridge. Walking along that ridge top I recalled how in one of John Green's books there was something written that Native Americans had said about where to be and not to be in regards to the Sasquatch.

What was said was never to be in the valleys at night, and only travel on ridge tops. I thought to myself why would this be? And I looked around as we walked and I realized that from a military perspective why this would be a good place to move, and what came to my mind made a lot of sense. Firstly, often times the tops of ridges are less covered in thick brush and more easily traveled. Secondly, they allow better vision of an area to navigate directionally. On top of a ridge you can see if danger is approaching in the sense of predators, allowing four directions of easy escape. Also, gravity allows faster movement in escaping danger going downhill in the opposite direction from approaching danger, so ridge tops made excellent places to be for several reasons.

We eventually reached the interior of the watershed, and it was a fascinating area. It was mostly bedrock covered with foliage. There were small channels of water coming from a number of locations where water collected on the surrounding hill sides, finally converging to make the Washougal River. The water in these small channels was about six to ten feet wide and only a few inches deep, and these channels were the best way to get around in this giant "bowl", so we walked through the channels toward the far side heading in the direction of a saddle that we felt was a crossing area for the creatures from this river system to an adjacent river system.

Once we reached the base of the ridge we knew led to the saddle, we started our
climb. This was now very difficult as the way up was steep and covered in thick brush. We stopped for breaks once in a while as it was a difficult trek up this ridge, and at one point I took my map out to see where we were. This was approximately the 2200 foot level in elevation, and after looking over the map to get our bearings, I noticed something unusual.

Having grown up in and around the forests of the Pacific Northwest, I was very familiar with what nature does to the vegetation there—and also the various things wild animals do to trees and other types of vegetation. What I saw was unlike anything I had ever seen before: A Douglas fir tree snapped in half. Now, this being the middle of July, with the weather having been warm and very summer-like for weeks, there had been no storms of any kind that would cause damage to trees. The break had been done recently, and it was located in a closed canopy with larger trees providing protection from possible wind or storms.

(First snapped tree I found in the Washougal River watershed)

I measured from the ground up to the break: It measured eight feet one inch. I closely examined the trunk and bark for any sign of bear or any other animal, or other possible reasons causing this damage. There were none, the bark surrounding the break, and the rest of the tree had no markings at all indicating what had caused this.

The tree was located on a small ridge that ran parallel to the large ridge we were climbing up, and ran in a northerly direction. I walked along this to see if there was anything that might provide a clue as to how this tree became broken this way, and I was astonished by what I found.

I called for Jack to come and look with me and we saw another tree of the same approximate size, approximately one hundred yards north of the first one, snapped exactly like the first one. I went to it and measured the height of the break as I had done with the first one, it too measured close to eight feet above the ground.

(Same tree wider perspective shot)

I continued along this small ridge to the north and found eleven more trees broken the exact same way and each one was just within sight of the previous one. I remembered Bob Titmus showing me dozens of snapped alder limbs he collected in the Bluff Creek area not far from where Roger Patterson had filmed a Sasquatch in 1967. I did not think much of them at the time as I had seen this often as the result of weather.

I did however keep in the back of my mind what Titmus talked about with the broken limbs he showed me, just in case I began finding similar things in my field locations.

This, however, was far from what Titmus had shown me, and was far more impressive. I had no idea at the time what to think, and noted my measurements in my pocket notebook and took plenty of photographs. Jack and I had spent a lot of time inspecting the damaged trees and when I checked the time, knew we had to retrace our path so we would be able to get out of the area before dark. So we left, deciding to return and look more and to attempt to continue to the saddle area as we felt that was a major site of activity and would potentially provide much more evidence.

When I had my film developed, I had shown the pictures of the tree breaks to PCSIT staff and board members, and field team members. Several of our members were friends belonging to the Klamath Tribe in southern Oregon. They told me these trees had indeed been damaged by Sasquatches. There were a couple of possible reasons for the markings. One could be territorial markings—likely the group leader (alpha) was moving on to the next foraging/hunting area and left markings for the rest of the group to follow.

This got me thinking about the larger picture of their movements throughout their range territory. Subsequent investigations would greatly change my thinking about the size of their ranges and movement patterns. As time would prove, with little or no human disturbances in a group of Sasquatch range would make it possible to accurately predict within a thirty day window where in an area a group would be located.

I spent a lot of time after this find in planning for the Eagles Cliff event. We decided to schedule the two day event for a time between the fourth of July and Labor Day holidays at the beginning of August. This was determined to be a time when people would not have much to do between the two holidays, leaving a window to maximize attendance.

The Eagles Cliff campground was filled to capacity during this time period, and we felt an event there at this time would get a lot of participation. We scheduled the event for the first weekend of August 1992.

I called René Dahinden hoping he would attend and possibly speak. When I explained what we were planning he became very enthusiastic about it and agreed to give a talk and bring slides from the Patterson film and some other items. I knew René would do a great job of a slide show/discussion on the topic of Bigfoot and that he was the star attraction for the event.

René's talk would be during the evening of the first day, so what to have to fill the rest of the time? I discussed this with all the PCSIT members in a general meeting following my call to Dahinden. We decided to make displays of evidence, and maps of areas the PCSIT was working in the field.

I worked for a steel manufacturer and many of my co-workers/friends were welders and could fabricate almost anything from steel. I got permission to take six or seven steel 55 gallon drums that contained shot for the blast machine at work, and Dan Hoggatt and Jayson Burch started work making them into BBQs. They cut them in half and attached screens. We had 13 of these for the event and they would be filled half way with sand with briquettes placed on top which acted as a heat barrier.

The BBQs would be placed on wooden saw horses, and I purchased hamburger and hot dogs to give away to attendees of the event. Our members would cook the hamburgers and hot dogs and chat with guests, and the campers could bring potluck food items if they wished.

Kevin had a stage near the campground, and we planned to use this during the day to show movies with Bigfoot themes on a television for anyone wishing to watch. Kevin had planned to have a bluegrass band play during the first day also but was unable to arrange for the band to arrive that weekend.

Kevin hoped the event would boost his summer sales as he made most of his income seasonally, and in return the PCSIT staff was provided free of charge the use of his largest two story cabin to stay in during the event.

Jack and I spent many exhausting hours over days planning for every detail, and when the day came Dan Hoggatt transported all the equipment to Eagles Cliff in a large truck, and Jack and myself were the last to arrive attending to last minute details.

(Poster advertising the event)

Jack had reached out to news agencies for advertising the event, and I had paid a talented co-worker, Dave Ward, to make a poster for the event. Dave subsequently produced a number of additional Bigfoot-related cartoons for me, (I paid him $25 dollars for each one), and Jack took the one advertising the event to make posters we planned to post in places along highways leading to Eagles Cliff, hoping to attract many people to the event. However, as luck would have it, the printing was not ready in time for the event, and we only had two posters to use. We placed them at strategic locations hoping they would be enough.

Jack and I arrived at approximately 10 pm the night before the start of the event at Eagles Cliff. We had been very busy making certain all the last minute details were taken care of in Vancouver and by the time we arrived I was exhausted. All the PCSIT members of the Board of Directors and staff members had arrived earlier that day setting up all the displays, tables and BBQs, and coordinating with Kevin.

René Dahinden had arrived earlier that day and had helped the PCSIT members set up. He came to me and we chatted for a bit. He then told me I looked tired and should get some sleep before the next day! So I went and fell asleep.

The big cabin Kevin had provided for the use of the PCSIT easily accommodated a large group of people. One of our staff members showed me to a room by that had been saved for my use. I was asleep moments later, hoping everything would be ready for the following day.

Early the next morning the cabin was a beehive of activity, with everyone preparing themselves for the day's tasks that lay ahead. Breakfast was being prepared in the cabin's kitchen and I could hear René Dahinden joking with the people making it.

(We had a big day ahead and coffee was a must!)

Someone came to my room with a cup of coffee and said René and everyone else was waiting for me to join them for breakfast, so I got dressed and joined them in the kitchen.

The atmosphere was very energetic and jovial, and to this day that morning remains a very fond memory. René Dahinden previously never seemed too interested in events such as this, but this time he seemed totally invested in what was about to take place.

We all talked about everything that was planned for the following 2 days, and I felt confident that guests would enjoy themselves. It was not long before we saw out the windows that people began arriving. I did not think guests would arrive so early, but they did and everyone hurried to get to their places.

In Search of the Unknown

(Eagles Cliff store and campground near Cougar Washington)

We had planned to have our members situated around the area to be able to answer questions and to chat with visitors. The event was as much a showcase of our members as any of the displays, in hopes of connecting with the public.

People began arriving around 8 am that first morning, and while we had no idea how many people were there at any one time, by the early afternoon Kevin came to me saying that his store had completely sold out everything and that he had to drive his flatbed truck to Vancouver to completely restock! He was both elated because he had sold in a few hours what normally took six months to sell, and also worried that he was out of everything with more people were coming all the time. He said he would be back as soon as he could, as he wanted to watch René's presentation.

Jack and I were happy that Kevin was having such a good day. This had been one of our goals in having the event at Eagles Cliff.

(Jayson Burch and Dan Hoggatt making burgers)

Representatives of various media outlets began arriving about this time also. Numerous newspaper and radio reporters and an ABC television affiliate from Spokane came. The media people joined in the festivities throughout the day, and did not stand out in the crowd until they started working as we started the presentations that evening.

Shortly after Kevin left to get more supplies, the PCSIT staff members set up the BBQs and began cooking. Now we wondered how successful the potluck participation would be. To all of our surprise, it seemed like everyone brought food. Soon after the hamburgers and hot dogs started being cooked the tables set up for food began being filled by campers and even day-guests brought food. The food part was an overwhelming success, as there was far more contributed than we could ever have hoped for.

(Attendees looking at some of the displays)

There were many people looking at the various displays we had set up. Everyone talked with our staff members, asked good questions and seemed to enjoy themselves a great deal. Myself and several senior members monitored the event, and it seemed like a huge success.

About six o'clock that first evening Jack told me we should start the presentations. I agreed as everyone seemed to be about finished eating and was beginning to gather at the stage waiting to hear René Dahinden speak.

Jack had some of our members clear the stage of the television and chairs that had been set up there. René's screen was readied for his slide presentation and Jack then got on the stage and announced that the evenings presentation was about to begin.

(Jack Livingston talking to the gathering crowd and making my introduction)

Jack started by thanking everyone for coming and hoped everyone had enjoyed themselves so far. He thanked our PCSIT members for all their work that day and then introduced me.

I entered the stage and gave a short talk again thanking everyone for coming. The PCSIT staff members had told me they did not want any spotlight and not to mention their names in public, so like Jack, I just thanked them for their hard work that day. I then gave an introduction for René Dahinden.

By this time, all the media representatives were in place to record the presentations. I do not know when they had set up as my focus was on having everything ready for the group in attendance.

(William Jevning talking to the audience and making Rene' Dahinden's introduction)

Prior to the presentations there was a near incident. I had been in the cabin talking with several members of the PCSIT Board of Directors and René Dahinden, when someone noticed that Datus Perry had arrived. Datus coming to enjoy the event was of course not a problem, but he apparently was in a very agitated state.

Some of our members had encountered Datus setting up his own displays, and he was angry that he had not been asked to be a guest speaker. He planned to be heard at our event even if he was not welcome to speak.

(Rene' Dahinden talks to the audience prior to his slide presentation)

We did not want Datus to speak as he had some very strange ideas about the Sasquatch, and did not have anything valid to present. I was going to go and talk to Datus, but René Dahinden stepped in and said that he did not want me in a confrontation with Datus, especially with so much media presence. He said that my image was important and that I should not be the one to get into any argument with Datus, so he would take care of the situation himself. René, Kevin and several senior PCSIT board members went to talk to Datus, who apparently agreed with what they told him. René later explained that he told Datus this was our event and that it was on private property, so he could not just do whatever he wanted. René said they told him he could stay and enjoy the event if he behaved himself, but that he was not allowed to speak or display his items. (As I mentioned earlier in this writing the displays Datus had amounted only to his wooden Bigfoot and a flip chart containing items from John Green's books.)

(Rene' Dahinden during his slide presentation)

The evening went very well, and even Datus enjoyed himself. After René had completed his slide presentation and talk, the crowd began to disperse after many people gathered around the stage to ask René questions. René seemed to be completely in his element.

No one had counted how many people were in attendance that day, it did not occur to any of us to record how many people came. Kevin had a guest book inside his store. Only a fraction of our guests had set foot in his store, but Kevin told me that evening that approximately 460 people had signed his guest book that day. His estimate was that at least two thousand people had come through the event that first day. He said he felt that was a very conservative estimate based on his knowledge of his store and campground. I was amazed, especially given that we were unable to advertise the event as planned.

(The ABC affiliate from Spokane Washington filming Rene' Dahinden during his presentation to the audience)

After René was finished chatting with people at the stage and the guests had mostly left, we all went to the cabin to unwind and discuss the day's events. To my surprise, all the media people followed us to the cabin and wanted to join us.

René and I had been walking together and chatting and we were the first to reach the cabin. René stopped at the door, telling all the media reporters that if they wanted to join us they had to check all recording devices at the door. They all agreed and we all went in.

We relaxed, drank cold beer and joked a lot and discussed how well things had gone the first day of the event. The reporters blended in like they were members of our group and joked right along with the rest of us. It was a very fun evening. They did ask plenty of questions about the subject of the Sasquatch and we did our best to answer them. Overall we made a very good impression on the reporters that night.

(William Jevning, Rene' Dahinden and Jack Livingston talk with reporters after the first day of the event)

The evening's festivities went to late hours, and eventually we all called it a night and went to sleep to be ready for the next day.

In the morning the cabin was again a beehive of activity. The main difference was that we had planned no presentations for the second day. Instead we intended to provide coffee and pastry for guests, and to interact with people in hope of collecting witness accounts.

The reporters had also stayed the night and wanted interviews with René, which he granted. He told me that he wanted our second day to be a success. Knowing we intended to interact with witnesses, he offered to take all the reporters to the Ape Caves, where they could ask him questions, leaving us privacy.

It was not long that second morning before numerous witnesses of Sasquatch encounters began coming forward to our members and discussing what they had seen.

We set up the area surrounding Kevin's store with private spaces. We placed chairs and tables so that small groups could gather with relative privacy to talk about events that they may not wish to share publicly. All these small group locations filled quickly with our members stationed at each one and coffee and pastry to enjoy.

We gathered many encounter stories—many were like so many others written about and all previously untold. We would later incorporate what we had gathered into the building picture of the creature's movements in the south Mt. St. Helens region. It became very useful in our field work.

At one point Don Turner our Director of Photography came to me and quietly told me he had been talking with someone I needed to talk to. This man was very reticent to talk to anyone about what he had encountered but had been listening to me and felt he could talk to me.

Don pointed to a man near the group where he had been having a discussion. Don then went over to the man and brought him to my location introducing him. For privacy reasons I will call him "John" in this writing. John was a large man in his 50s. Later John told me that he was 6'6" tall and weighed around 300 pounds, and had been a biker most of his life. He had a couple encounters with creatures, but had never told anyone but those close to him.

He told me his first encounter had occurred many years earlier in his youth, he and a fellow member of the motorcycle club he belonged to had decided to go to southern Oregon to look for the Port Orford meteor. This meteor had been reported to have crashed somewhere in the area east of Port Orford Oregon and was said to be worth approximately one hundred thousand dollars John said, so he and his friend had decided to look for it in hopes of collecting the reward for finding it.

He told me that they rode their motorcycles to a location near a lake, and decided to camp for the night and start searching the following day. That night while sitting near their camp fire they began to hear strange screams from the nearby ridge top. The screams seemed to be moving down the ridge in their direction, and they thought someone was messing with them. These were not men to mess around with, and John told me in his youth had done some very bad things and had a reputation for it.

Both men had .30-30 caliber rifles, and shouted that whoever was there had better show themselves or be shot. They got no answer to their challenge. Not long after they had called out to the visitor, this huge hair covered creature burst out from the nearby tree line on two legs like a man and ran toward them in an apparent rage. It stopped at the edge of the light from the fire and grasped two nearby sapling trees, one on each side of it. The creature then proceeded to violently thrash the saplings about in an apparent display of anger, growling at the two men.

Without hesitation both men simultaneously shot the creature in the chest, all it did was screech loudly and run off into the darkness. Both men were extremely frightened and immediately got on their motorcycles and quickly left the area, not looking back and not returning.

He told me later when they stopped many miles from the area they both felt as if the creature had intended to eat them. He did not know why they felt like that but the impression had been very clear to both men. They vowed never to tell another soul what happened.

The second incident happened in the mid-1970s near Larch Mountain south of Mt.
St. Helens. John said he and a friend had been in the area one night poaching deer. They had been using a spotlight while driving slowly in the mountainous area near Larch Mountain in his Jeep CJ5, and had spotted a deer on the switchback above them. They shot the deer and saw it immediately fall, feeling confident they had killed the animal.

They drove to the location where the deer had fallen, and found something they had not expected. There were a few inches of snow on the ground. The spot where the deer had fallen was covered in blood. The deer was no longer there however, there was clear evidence something had dragged it away. They followed the drag markings in the snow with the spotlight to where they saw a huge man-like hair covered creature grasping the dead animal by its neck dragging it with one hand.

He said the creature never turned to acknowledge their presence and never changed its pace. It just kept dragging the dead deer up the road. Neither man had any interest in trying to wrest the deer from the creature and were quite frightened by its size and appearance, and decided quickly to let the creature have the deer and quickly leave the area. He said that was the last time he ever went poaching.

John told me he liked what he heard me talk about in regards to the creatures, and had extensive knowledge of the regions south of Mt. St. Helens and knew of numerous Sasquatch encounters by local residents, and said he would help me in any way he could if I was interested. I told him I would very much appreciate any help he could provide, and went on to have some adventures with John.

By the afternoon of that second day at Eagles Cliff, most of the people who had come to the event had left and the PCSIT staff was busy packing up all the equipment and displays.

Kevin could not thank us enough for such a successful two day event, and was looking forward to what we could do the following year. We seriously discussed holding an annual event there. The next one would be the "second annual" BBQ and Potluck. Jack suggested my impromptu talk be the first presentation, and from a marketing perspective the second annual event sounded better, so we followed his advice.

René returned after spending a few hours at the Ape Cave with the various reporters and several senior Board members and I sat in René's homemade RV chatting with him about the event. René had been very enthusiastic about what we had put on that weekend, and told us that nothing like it had ever been done in the history of the issue of Bigfoot, and that we had a lot to be proud of. René was very impressed with the large crowd we had attracted and the perspective in which we had presented the subject of Bigfoot.

Jack and I had talked about some of René's interesting possessions and how he never had much money. I thought he should be selling some of the unique things he owned. Among them was the copyright to Roger Patterson's 1966 Bigfoot book. He also owned footprint casts Patterson and Gimlin made after famously filming the Sasquatch in Northern California in 1967. The two men cast three of the creature's footprints after it left the area. One of the casts had broken and they were left with one left and one right foot cast. René had shown these to me years before during one of my visits to his home. I thought he should make molds and sell copies like Grover Krantz did with footprint casts others had made.

I felt he deserved to have some income from his many years of hard work, so we discussed these ideas and René agreed to hear us out. We presented him with some of the ideas for his consideration. He then said he needed to start his trip to British Columbia before it got late, and we said our goodbyes and he left.

The staff had already left with all the materials we had brought and like the night before the event was to begin, Jack and I were the last to arrive and also the last to leave.

On the drive to Vancouver we talked about the weekend's event, but my mind was on the future and what may have come from the event we put on, and as always the next step in my field work and how this may have helped our efforts. A couple weeks after the Eagles Cliff event, René called me. He didn't seem to have any reason for the call except to chat, and during the conversation he mentioned that his birthday was coming up in a couple days. I asked him if he had plans to celebrate and he said no. He was just puttering around the house and he sounded sad that no one was doing anything for his birthday.

I immediately had an idea, and said that Jack and I should come up to visit and we could discuss ideas we had for him selling things and re-publishing Paterson's and Don Hunter's books. He perked up immediately and agreed that we should come up.

I called Jack after I finished talking to René and told him we should do a surprise birthday get together for René that he had no one to spend the day with. Jack immediately agreed and I called some of the field team members and staff of the PCSIT to see who wanted to go with us.

None of us had any idea what sort of gift to get so what we decided was that our group celebrating his birthday was the gift, and we bought gag gifts since René had a great sense of humor and someone made him a birthday cake.

(Rene' Dahinden during our surprise birthday party we gave him)

When the day arrived we traveled with three caravans and made the trip to Richmond, British Columbia. Jack and I pulled in first as René was expecting the two of us. We were about ten minutes ahead of everyone else. Jack had brought his two young boys, Jack and Jon and René immediately smiled liking children so much.

It was a total surprise when the rest of the group arrived, and René told us that no one had ever made a big deal of his birthday before and he was both surprised and very happy we had all come.

We all spent the night camping in tents outside René's home. We talked long into the evening, sharing dinner and stories. The following day as we prepared to drive back to Vancouver and Washington and René once again thanked us all for coming, saying that we had made his birthday something to remember.

(This is the 1968 Volkswagen van I purchased from Jack)

One evening after we returned from British Columbia, Jack and I were talking at another friend's home, and the subject of cars came up. I had seen an old Volkswagen van parked in the yard of Jacks parent's home with grass growing up around it. I asked Jack about it. He said he owned that and other than needing new brakes and the fact that the floor boards in the front were rusted through, it ran fine.

I asked him if he had plans for the vehicle, and he said he didn't have the money to do anything with it as he didn't have a job at the time. I asked him if he would be willing to sell it, and he said he was. We agreed on the price of three hundred dollars and I had the van towed to a mechanics shop to have the brakes replaced and the floorboards fiber glassed.

(I used this van often during my field expeditions, it was very good for such outings, this was a trip to the Mt. Adams Washington region)

About a week later I picked the repaired van up, and it was like new. The total investment of eight hundred dollars proved to be one of the best vehicles I have ever owned to take into most conditions during my field work the years of the early 1990s.

John the man we had met at the Eagles Cliff event had been helping us search the upper Washougal River and Yacolt areas. Having lived in the region many years he had extensive knowledge of the people and sightings of the creatures in the area.

He took us to a location where he said years earlier he had discovered an old gold prospect, and there had also been numerous Sasquatch encounters which had caused the original gold prospectors to abandon the claim. He wanted us to go up the mountain in hopes of locating the claim and said we might find evidence of the creatures in the area there. He was unable to hike due to suffering advanced osteoporosis, but insisted we take along a close friend of his.

His friend going along was his insurance that should we find the gold claim we would not say we had not then later claim it as our own. We had no interest in the claim, but agreed since his friend was acting as a guide to the place he said had seen numerous Sasquatch encounters.

Jack and I were both in our early 30s at the time and in pretty good physical condition to make difficult treks cross country through thick vegetation and forest while climbing steep mountain sides. John's friend was 55 years old, overweight, and obviously not in good physical condition for the journey. We immediately knew this was going to be a slow trip up the mountain.

To begin the trip we had to cross the Washougal River on foot. I located a place where we could cross with relative ease. Once across the other side there was a pool of water we also had to cross that was almost waist deep.

The water was still and calm, but the bottom was bedrock and covered with algae and slippery. I would be the first across to test the footing and show Jack and John's friend where and how to get across.

One thing I was apprehensive about was that John's friend insisted that he carry all the gear in his back pack, including the only radio we had to stay in communication with John in case anything happened during the day up the mountain. He also carried a double barreled twelve gauge shotgun in case we ran into hostile Sasquatches.

In Search of the Unknown

Jack and I knew well the need to have the proper attire for the journey up the mountain, and had dressed appropriately. John's friend, however, was ill prepared but my thoughts were that he was a grown man and should know better. His footwear caused me most concern. He was wearing a pair of what were called "moon boots", basically little more than high topped house slippers. I knew was going to have trouble as we made our way in this hostile environment, as he would have absolutely no support for his feet.

John also insisted that his friend carry all the equipment because he had supplied it, while Jack and I carried cameras and light items such as water and a snack for later in the day. We had little choice but to agree and see what happened as we went.

I found a long tree branch to use as a third leg to help maintain stability on the slippery surface of the bottom of the pool. Moving at a slow but steady pace, I made my way across the pool which was about thirty feet across. Once across, I shoved the branch I had successfully used across the water to Jack. I instructed them to have John's friend cross next so there would be one of us on each side of him in case he needed our help.

I told him to do exactly as I had done, using the branch to help maintain balance and to slowly shuffle his feet along the bottom of the pool. I said that if he tried picking his feet up to walk across the pool he was going to slip and fall in and that he needed to use my method.

Halfway across the pool, we could see he was losing balance for a variety of reasons. Wearing the totally wrong kind of footwear, not being strong enough to maintain his balance and he was taking steps by lifting his feet instead of shuffling or sliding his feet along the slippery rock. The inevitable happened in what seemed like slow motion. He went over backwards, ending up completely submerged in the water. All the equipment was completely soaked.

Jack and I looked at each other in disgust. We knew from the start that bringing along someone so completely unprepared for this kind of journey was going to be a huge problem. In later years we often shared a good laugh at the look on John's friends face as he slowly went over and under the water. In retrospect it was one of those events that stay in your memory and was actually very humorous but not at the moment.

Now at the beginning of the day we had no radio in the event of an emergency, or if anything happened and we needed to call John. I was in charge of the expedition and decided to press on. John's friend had insisted on putting himself in the position he was in, and was now going to have to deal with it. Once we got him up to the surface and on the other side of the pool Jack crossed successfully and we started up the mountain.

The underbrush was quite thick and difficult to move through as much of it was face height. One had to take care not to get a face full of twigs and other materials that were not pleasant. John's friend had a lot of difficulty with this.

We reached a place in elevation that finally brought us out of the thickest of the undergrowth, and the trek became easier but not less steep. Eventually John's friend could no longer climb up the slope. His moon boots provided little traction and frequently slid off his feet.

In Search of the Unknown

We had not located the gold claim, it became apparent that our guide had no idea where it was located and there was no sign of Sasquatch activity. I knew that due to the slow pace we had kept, we were well behind schedule and in order to make it out of the area before darkness set in we would have to start back to where John was waiting now.

John's friend was also unable to go any further and he was too large for Jack and me to carry down the mountain. I told them that we had two choices: We were going to have to attempt to make the descent and get out before night set in, or we were going to have to leave John's friend there and Jack and I could get to John and he could go for help to get his friend down off the mountain.

John's friend did not want to be left there as he was afraid of the Sasquatches in the area, so said he was going to try his hardest to get out with us. There were numerous times he needed to stop to rest and was in tears from pain and exhaustion.

I tried my best to find the shortest route and clearest of brush down the mountain. The best way was in a seasonal creek, dry this time of year. The trouble with the creek bed however was it contained many devils club plants that averaged about six or more feet in height, and we were able to navigate around these except John's friend seemed to be unable to avoid some of them getting thorns puncturing his arms and legs, the entire day became something of a nightmare for him.

Just at dusk we arrived back at the pool of water, this time we helped John's friend across the pool one of us on each side of him holding his arms so we could quickly cross the water.

The trip had been unsuccessful in terms of finding anything useful, but valuable in my making the decision to take people along on my terms only from that day forward. Safety of expedition members and the mission itself being compromised allowing others to dictate who and how an expedition I was leading would be carried out was completely unacceptable after that day.

John even said afterward that he should have trusted my judgement and that I was right about how the expedition should have been carried out, and that any future trips he would follow my advice to the letter.

Not long after that day John called me to tell me that a friend of his who was the owner of a chain of music stores who was wealthy wanted to produce a documentary film. The film was going to have several sections, one portion would cover the gold prospecting in the area, and in particular the claim we had tried to locate with his friend. Another section would be about Bigfoot and Jack and I would be involved in that portion, there were other parts covering the Yacolt burn where a large forest fire had devastated the region nearby earlier in the century.

He said his friend would discuss the details with us but that he would pay each of us one hundred thousand dollars for our participation which I immediately thought would put my own field work in a far better position enabling me to get equipment and be out much more often than I was on my limited income now.

Jack and I had hats and shirts produced with the PCSIT name and logo on them for recognition purposes on film. Plans were made to do some preliminary filming of the area and a walk through to make the plan on the actual filming of the area where we had taken John's friend previously. I had explained the difficulties of the area and we decided we needed to look at it to decide of filming there was practical.

In Search of the Unknown

We had planned to spend a weekend near the place we crossed the Washougal River on the ill-fated trip with John's friend, This was also in the general area where we had found the snapped trees on another expedition into the watershed, so seemed like the best entry point and I thought we could get a general look at the prospective filming areas from there.

The friend of John's I will call the investor seemed like he was physically capable of trekking the area with us and he had brought along his brother who was also investing in the film, and was an Oregon State trooper so I thought these men were legitimate and felt that this may be w worthwhile venture.

On the first day after we established a base camp, we readied ourselves for the trip and the investor carried a professional film camera, he and his brother the State Trooper carried light items the same as jack and me, so I felt this would be a good day. We easily crossed the river at the pool mentioned earlier and we completed a circular trip around the north side of the watershed region, the investor filming as we moved along and I talked about the area and the Sasquatch encounters there and the surrounding region.

We left the area later in the day, the investor gathering a lot of footage that day and he said it had been a very good day. He wanted to go farther into the interior of the area the next day to where we had found the snapped trees, I agreed as this would be interesting on film and that we might find more evidence in that area.

When we arrived at the base camp, we sat down with John and filled him in on the day's work. We were pretty tired and it was a warm day and a cold beer seemed to be in order. We had not brought any with us not knowing what we would be doing, but since the day's work was done we felt we could relax for the rest of the evening. Jack and I took my VW van to the nearest store about ten miles from our location.

We had a real scare driving down off the mountain; it was an eight mile drive down with many switch backs down the steep slope of the mountains there. About a mile or so driving down the mountain the vans brakes went out, and Jack and I had only the option of jumping from the moving vehicle before it went over the next hairpin turn and over into a thousand foot canyon. Jack said "hey, shove it into second gear!" which I quickly did and the van stayed at a constant rate of speed slowly until we reached flat ground at the bottom of the mountain.

I had not been that frightened in a long time, and we stopped when the van finally was out of momentum. We did not have cell phones back then and had to either walk to the nearest phone still some miles away or wait for someone to hopefully come along we could get a ride with.

We sat there for a time talking and deciding what we would do next when here came John in his truck. At first when we saw him we thought perhaps he decided he too wanted something from the country store we were driving to but when he stopped he had something quite different to tell us.

We had told him our harrowing trip down the mountain and he said he was glad we made it ok, and that he would tow the van to a friend's shop for repairs. I asked what about the filming? And he said that's why he came to find us, he found out something we needed to hear.

When we left camp for the store, the investor had revealed the true reason behind his interest in making a film. It had nothing to do with making a documentary of the various subjects we had been told. The sole purpose for his wanting us to take him and his brother into that area was for me to lead him to a Sasquatch that he would film, and then everyone would be cut out of anything gained from the film. Essentially he was using Jack and me to become the next Roger Patterson then he was going to disavow our having any interest in the film what so ever.

I was livid, and said had I known what he was up to I would have never taken them in there. John agreed and was lied to also about the pretext of the project. We went back to the base camp and I planned to confront those two gentlemen about what I knew, but apparently John's conversation with them had been enough for them to pack up quickly and they abandoned the camp.

We packed up our equipment and towed my van to the shop, this was one of fortunately few very distasteful encounters I have had over the years with people attempting to use my knowledge for personal gain, unfortunately it would not be the last.
I called René Dahinden and told him what had transpired with the two men trying to use and cheat us, and he told me to always be careful that the world was full of people just like those two men. He said many would try to use what I knew then cheat me out of any gains from these ventures, he said I was better off keeping ventures like this to ourselves, getting involved with outsiders was always a big risk and often detrimental to us.

He told me he was planning a trip soon to Northern California and would call me when he was on his way back, and that maybe we could get together and talk, I said I was looking forward to it and would fill him in on what I had been doing in the field.

A few weeks later I received a call from René, he asked me if I wanted to meet with him the next day in the town of Estacada Oregon for breakfast at the Safari Club. I said sure that would be fine, and called Jack but he was going to be away on business so called my old friend Carlo Sposito if he wanted to join René and I for breakfast the next morning in Estacada. He said sure it would be great to see René again, so we made plans to drive together and I would fill him in on the work I had been doing.

The following morning we met René at the Safari Club in Estacada, and got a booth so we could talk more privately. René was always full of humorous stories and we thoroughly enjoyed the breakfast get together. Since we were in Estacada, I asked René about a small piece that was in one of John Green's books about the logger that had watched three Sasquatches back in 1967 in this region, the two adults stacking rocks. I wanted to know what he thought about that story.

He said they had no reason to doubt the man's story and there had been little in the way of verifying the account, but that the man's story seemed legitimate. When Carlo excused himself to use the restroom, René told me to call him in a few days about the rock piles, and to keep it between the two of us. When Carlo returned to the table, René asked us if we would like to see where the logger had seen the Sasquatches. We asked how far the place was, and René said about an hour drive from town so we agreed to follow him to the place.

We had spent the entire day going to other places where René talked about people seeing Sasquatches, and that after giving us a witness tour of the area he would take us to the place the logger had his encounter. It took much longer than we had anticipated that day with René, but as always having none less that René Dahinden giving us a tour was well worth the time spent.

(This is one of my pictures of the first site Rene' Dahinden took us to)

Right at dusk that evening we reached the place where the logger ad seen the three creatures, and René pointed to where a small trail was and told us to follow that and we would find the rock stacks still there. He wanted to go on as he wished to look over some other part of the region and camp for the night and head back to British Columbia the following morning. We parted company with René and walked along the path that René had shown us.

We arrived at a rocky area and indeed did find rocks stacked, but it was getting dark now and we could barely see them. I marked the location on my map intending to return in the morning to take a good look.

(William Jevning sitting next to one of the stacks for scale, this is from site number 3)

I returned the following morning and again found the stacks of rocks, each one three or four rocks, and each rock was pretty large. I did not know what to think as I walked among the stacks, and even found nearby the large hole the logger said the male Sasquatch had dug and extracted a nest of rodents from. The hole measured approximately six feet across and six feet in depth, quite a hole for a few rodents. As agreed, a few days later I called René and asked him about the rock stacks and what he really thought about them.

René told me that after interviewing the logger, he began doing his own searches in various places in both Washington and Oregon and had located similar places to the one the logger had witnessed. These sites were completely unknown to anyone but himself.

(This is a rock pile from one of the sites Rene' Dahinden told me how to find)

René told me he wanted me to go to the sites he had discovered throughout the States of Oregon and Washington and fully photograph and document them, and that we would discuss all this at a later time. I agreed and he gave me instructions on how to locate each of the sites.

I made plans to travel to the first site that was known and begin there, as basically a testing step in my documentation process, I took some of the PCSIT field staff with me and went to the Estacada site. We set up base camp and I came down with the flu, but being very stubborn went ahead with my plans to document what was there.

In the months ahead I went to the other locations René had given me alone, there were none total in both states. Most contained rock stacks, some had rock piles and still others contained both. To this day I continue to learn of more of these as far north as northern British Columbia.

(Rock stacks from site number 1)

I went for a weekend visit to René's home in Richmond British Columbia after photographing and making detailed notes about each one of these stacks and piles and the hole and a partially dug second hole at another region I found. I brought along the slides and photographs I took for René and I to look at while discussing them.

The main feature of every one of the sites except the first was that they are all completely inaccessible by any means except cross country travel by foot.

None are close even remotely to any road, trail or any form of human access way, and seem to be purposely placed in difficult to reach locations. At the time neither René nor I knew what exactly to make of them or why the creatures would even bother with doing such things. We decided since such little information was yet available that we would keep an open mind to this and wait for a possible future find that might present clues to this behavior.

(William Jevning standing in the hole the logger witnesses a Sasquatch dig)

In 2003 when I traveled to the Klamath Tribe in Southern Oregon, I was doing research for my first book Notes From the Field and brought the topic of the stacked rocks up with Gerald Skelton then the director of the tribal cultural committee and showed him some of my photographs, he told me that Native peoples would routinely stack rocks during prayers, but they were always very small and using small rock. He was interested but did not have any idea what they represented, but would ask around with some of the tribal elders, perhaps they knew about them.

To this day they remain a mystery, I am still waiting for a find that may reveal more about the enigmatic stacks and piles and continue to keep the locations of these sites a closely guarded secret.

I continued conducting my searches throughout the vast region south of Mt. St. Helens, and interviewing witnesses. I started finding patterns to the movements of the one group of four creatures we had investigated during the Yacolt events of 1989, Over a decade of following their movements I was able to narrow down to within a thirty day window just where in the region they would be, unfortunately I seemed to always be just behind them and with such difficult terrain and time and financial restrictions I had to do the best I could given the circumstances, still I learned a great deal and knew I would be able to narrow the window of opportunity even more eventually.

Behaviorally the Sasquatch became very different than the picture the early investigators had of the creatures. Instead of the shy, solitary inhabitants of the forests the picture that emerged throughout hundreds of witness accounts I learned of were quite different.

For example, they often were seen as multiple individuals or groupings. Seldom solitary in reality, what I learned was they most often when seen as solitary individuals was only a small picture of a larger one that was happening in a given area, often with other identifiable individuals in relative close proximity to the "solitary" creature first reported. Even my own initial encounter was with two of the creatures, and two years previous to that encounter we had found the footprints of three of them. The Washington State Trooper Mark Pittenger saw all three of the creatures that were the same ones I had seen two of.

There were lots of witnesses such as one gentleman I spoke with one day. I had been at one of our PCSIT board member meetings in Skamania County at one of the board members' homes and after the meeting ended decided to take a walk. This was a very rural area with all farms on acreage. I walked down the road thinking and looking the area over when a man came driving to the end of his driveway. He stopped and we had a conversation and I told him what I was doing. He immediately told me that he had once been in the Marine Corps as an intelligence officer to give me a little idea of his back ground. He told me that a couple weeks ago while coming to this location at the end of his driveway to check the mail at his mailbox, he had seen what he thought at first to be someone walking along the tree line across the nearby field of his neighbor, but it turned out to be a huge man-like hair covered creature.
He watched it walk along the edge of the field, it apparently not caring that he was there watching it, and it eventually turned and walked out of sight into the forest.

He said he was stunned, people had mentioned there being such things in Skamania County but he thought they were just local myth until he saw one only a hundred feet away in broad daylight.

He told me there was more, that his next door neighbor had been having a lot of trouble with the creatures especially around deer season. His neighbor had a shed that he would hang his deer in to cure, and that several of the creatures had taken his game hanging in the shed on a couple occasions, and he finally got mad at this and closed the opening of the shed up and locked its door when he had game in it.

The last time they came after he had shut off access to the meat in the shed, his teen age daughter heard a loud ruckus one afternoon at the shed, she went out to see what was going on thinking someone was trying to break into it. They had thought people were stealing meat and had no knowledge of the creatures taking the game. His daughter saw three creatures violently trying to shake and enter the shed; she screamed and called for help. By the time some arrived the creatures had fled the location, but continued coming around each hunting season, but they had quit hanging game or any meat in the shed after that event and eventually the creatures stopped coming there.

This example could simply have been the creatures foraging for food, and they always acted aggressively toward the people living there and could be explained as acting that way over the food, but this became a common thread among the hundreds of witness I spoke with during my years in that region.

There were other things I began becoming aware of that I never considered even remotely possible in the subject of Bigfoot, and one was apparent governmental involvement.

Now I am certainly no conspiracy theorist by any means, but having seen the creatures with my own eyes, knew those that work for the government working routinely in the forests must know something of them.

I stopped one time at a ranger station in Northern Oregon while conducting some field work there and decided to ask the rangers there if they had ever seen anything or knew of any Sasquatch activity in their travels. There were two rangers present in the office that day. The first one smiled and said he did not. The second asked, "Why don't we tell him about the diary?" The first ranger said, "Sure, but that's all we know."

They had discovered a very old family Bible at a long abandoned farm/homestead. In the margins throughout the book along with the recorded births and deaths of family members, were notes about creatures. There was also information about the places they could allow children to play and where they absolutely could not, because the creatures were known to take children who were never seen again and presumed to have been eaten.

They did not have the book. It was sent to someone dealing with historical artifacts, and I never did learn of its location, but found that fascinating. I thanked the rangers and shared a little of what I knew of Sasquatch activity in their work region and I left.

I soon found myself involved in two separate incidents that leave me scratching my head to this day. The first happened in an area not far from Goldendale, Washington.

In 1994 the father of a co-worker had been scouting a favorite hunting area near Mt. Adams in southern Washington State with a friend. The area, not far from the town of Goldendale is heavily forested and sparse in human population, and the stories of Sasquatch being seen in that region go back to the earliest recorded history of that area.

The particular region the two men favored to hunt was closed to public access. To gain entry to the area, they built Baja's out of old Volkswagen beetles. These vehicles worked amazingly well to drive cross country through the forest there, and that's how they traversed that region, entering the road system going around closed gates or other obstructions.

They later told me that they had seen a forest service water truck not long before entering the forest. It was parked by the logging road with its occupant cutting the lower tree limbs near the road. They stopped and chatted with the young man for a few minutes but did not mention what they were doing in the area. They said he told them he was doing some fire prevention work and shortly afterward they parted company.

After watching the road for several hours for deer tracks, the two men came upon some very unusual footprints, and they realized right away that they must have been made by a Sasquatch. This area is warm and arid, and the logging roads are largely covered in a fine dust two to three inches thick. This dust is perfect for footprint impressions, and the men found hundreds of tracks along the right side of the logging road. They decided to get out of the area and report what they found.

When they drove out of the forest and onto the road leading to the highway, they saw a forest service pick up coming their way. They stopped to talk with its driver who identified himself as the area supervisor. They talked about hunting for a few minutes then told the forest service supervisor what they had found. The forest service officer, (I have since lost his name), immediately retrieved a metal case and took out what they said was a very expensive looking camera, and asked where they had found the footprints. When told, he said he was going to take pictures and document what they had seen and he drove off.

They decided that they needed to get someone out there to document what they had found. The one man whose son worked at the same company that I did at the time had been told by his son that I was a Sasquatch investigator. He contacted his son at work, who in turn reached me and asked if I would accompany his father to the location where they had found the footprints. I agreed and they were already enroute to meet me and go to the site.

It took about an hour to reach the location. By the time we arrived, it had been about 4 hours since they had found the footprints. We got in the Baja when we reached the point where they had entered the forest previously. (It was not street legal and had to be brought on a trailer). We retraced the same route they had taken earlier that day, and finally reached the place where they had found the first footprints as they had marked the place.

There were no footprints! Instead it was obvious what had happened, we saw the duel tire tracks of the water truck they had seen earlier, this is the kind of vehicle used to support forest firefighting effort. The operator of the truck had 'misted' the road only where the two men said they had seen footprints. (Misting is a heavy mist from a fire hose which saturates vegetation and uses minimal water. Very helpful in preventing the spread of fire.)

On this very powdery dust on the road it appeared as if no tracks of any kind had ever been there. It was actually a very clever way to erase them. But I had fought forest fire one summer and knew what to look for. Water evidence was obvious, so the footprints had been destroyed deliberately.

The two men had gotten the forest service supervisors name and he had told them they could get copies of the photographs he was going to take of the tracks. When we tried later to find this person, the two men and I were told on separate dates that no such person worked for the forest service, and no one had taken any photographs of supposed Bigfoot footprints!

The two gentlemen who took me to that area were very credible and not prone to making up stories. I have no doubt they were being completely honest about what they said they found. The remoteness and difficulty reaching the place they took me makes the story credible alone.

This event was really strange by itself, but soon after this a second equally mystifying event happened near Eagles Cliff store and campground.

A group of us had planned to do some field searches east of Eagles Cliff due to a couple recent Sasquatch sightings in the area. We stopped in at Kevin's store to see if he had heard anything new and his sister told us we had just missed a forest service biologist by about five minutes.

She said one of the local residents living near the Swift reservoir nearby had discovered a good line of Sasquatch footprints just up the road near a creek that morning. When word got out about the track find the forest service biologist shows up saying he was going to the location to document the tracks and see if they are real or hoaxed.

I was very familiar with the creek where she told us the tracks had been found, so we drove to the place. There were six of us in my van that day so we were able to conduct a search pattern on both sides of the creek. We swept the open areas, then looking inside the tree line for approximately a quarter mile in both directions. There were no sign of any footprints, not even the biologist which seemed very peculiar since this was the exact place the local resident that made the initial find had confirmed when we went back to the Eagles cliff store.

Something was wrong, and we knew Gene McKinney and his wife who were at the time employed by the forest service to do security for the forest service location. We drove to visit Gene and he said that he would call the main office to find out who this biologist was. When he got off the phone he looked puzzled, and said that they did not have a biologist on staff at that time and no biologist from another location had stopped in, which would have been protocol.

No such forest service biologist apparently existed, and what had happened to the Sasquatch footprints discovered that morning remain mysteries to this day. I had asked René Dahinden if he thought the Governments of the United States or Canada might have any interest or involvement in Sasquatch matters. His response was that they neither had the time or interest in getting involved, and that was the end of the matter.

In 1995 I was contacted by a representative from a television show called *Encounters*. They had heard of my involvement at Knott's Berry Farm in 1988 and asked me about recent Sasquatch activity and if I would be interested in participating in their show.
I told them I did have some recent witness encounters and that I would be happy to participate. At that time, Jack acting as the PCSIT marketing director took over dealing with the television show producers and apparently thinking he could get a large financial investment from them in the way of using a helicopter and other big expense items made the producers shy away from our involvement.

I later learned that this would not have been a good show to be associated with so was glad I did not become involved with it.

I was never interested in public exposure for the sake of exposure, so would never jump at any opportunity to be on television or radio. I always have been interested in the subject of Bigfoot to be presented in an even handed and credible light to the public. All too often television and radio does not portray the subject in any credible way so I am very choosey about my participation.

By the late 1990s the PCSIT as an organized effort was flagging. Many of our members had gone their separate ways, either no longer being able to devote time to the work or physically moving to regions or states far away, so as to make involvement impossible. By late 1998 those of us left had decided to close permanently the PCSIT doors, officially ending its existence from 1975 to 1998, the first non-profit organized effort to resolve the question of the Sasquatches existence covering a period of 23 years was over.

Even I was preparing to move to the south Puget Sound area of Washington to be near my family and help with my aging father. I had made my plans to move, quit my job and was ready to leave Vancouver when I had to stay an additional two weeks, so one of the former PCSIT board members let me store my belongings in his basement until I left and I took care of some business and worked a temporary job until I left. Tragically during this short delay his basement experienced flooding with a broken water pipe and most of my research and materials had been destroyed, when he called me I thought I might be able to salvage most of what I had there but this was not the case. I was sick to my stomach, so much work and careful documentation gone. Many of my photographs survived, but many were also lost.

I collected all I could and packed up and left Vancouver, knowing that I would have to pick up in old familiar regions and start over. I already had greatly expanded my research areas covering the entire west coast, and no longer having the structure of the PCSIT, I thought about what René always told me about doing my work alone.

After relocating to the Puget Sound area, it was not long before I found another job and took some time to re-establish myself. I traveled around to some of the old areas where Sasquatch activity had taken place many years earlier, but much had changed and I started looking at what had been reported around the region.

(William Jevning and my close friends Scott Martin and Bill Fox)

It did not take long before I started focusing on an area just outside Mt. Rainier National Park with a history of Sasquatch activity had been. My brother in law, Bill Fox, is a teacher and one of his students told him about being on a camping trip to this particular area and seeing a line of Sasquatch footprints near a creek.

I began working in that area, but it's very rocky and not conducive for animal tracks in most places, but kept searching anyway.

By 2001 when the events of 9/11 happened, I was covering many areas throughout the region and continued traveling once in a while to the Mt. St. Helens area, but now also searched the parts north of the mountain, and going to Oregon as well. The weekend following the 9/11 happenings, my buddies and I had planned a camping trip in the area where so much activity had been reported over many years and while intensely listening to the news on the radio that night, not knowing what might happen after those attacks we heard screaming begin.

Bill turned the radio down to hear the screams and even though we judged the source to be a quarter mile away it was clear it was no usual animal noise. I had heard this same vocalization before, most commonly south of the town of Yacolt during the long investigation we did there and was certain it was a Sasquatch.

This vocalization went on for about twenty minutes then we did not hear it again. On another camping trip on this particular mountain a couple months later, I found two snapped trees exactly like the ones we found in 1991 in the Washougal River watershed, so knew there was at least one of the creatures in the area. I spent a lot of time hiking and searching the area but the sign was old enough that any footprints were weathered away and no fresh sign visible.

(Stopping for a cold soda during a hike in the Glacier view wilderness near Mt. Rainier)

A couple years later we were camping on a creek just outside the Park boundary of Mt. Rainier and while sitting talking around our campfire one night we heard some very strange noises about one hundred feet above us on the steep ridge above our camp. There were two clear and distinct vocals from two sources. They seemed to be only about twenty or thirty feet apart and one was moving back and forth making a chattering sound, while the second source was making sounds very similar to a chimpanzee.
I had heard plenty of Sasquatch vocals over the years, but this was the weirdest if it was indeed from two Sasquatches. This went on for quite a while and the group of four or five on that trip sat there in amazement, none of us ever having heard anything quite like it. I climbed up to where I thought the sounds came from the next morning, but it was solid rock up there and only flattened moss so nothing definitive as to what made those sounds. That night remains a mystery.

I hadn't learned that my old friend René Dahinden had passed away until late in June 2001, I knew he had been sick but René was not one to talk about things like this not wanting sympathy. At least with me, we always just talked about what was going on with Bigfoot or jokes. I was very sad to learn he had died and sent a letter of condolences to his son, Erik. I told him that his father told me a number of times he regretted his choice of putting the Sasquatch before his family, but I never heard back from René's son. Thinking they were grieving I did not reach out again.

I later learned that René had been cremated, and they had left his remains in his old home made RV not caring what happened to them. Someone René had reselling his books had discovered his remains while cleaning out his truck. I was appalled when I learned this, but knew that his family reacted in accordance with how he treated them, so in the end was not surprised.

It was a shame, someone of his stature not being given the credit due him for starting the quest for something so important to be left ignored after his death.

I discovered that one of my old friends from Vancouver, Ed, was living in the town of Olympia when I was there one day and saw him. I stopped and chatted with him. He had recently left the Army and was married and working in town. We arranged for me to visit his home and meet his wife.

Ed was working selling computers back then and built them in his spare time at home. During my visit told me, "Will, you have to have one". I did not know the first thing about personal computers, so Ed showed me how to use one and gave me one he had built.

I was apprehensive about its value but thought I would play with it; maybe I could use it for something.

(William Jevning in the field Olympic Mountains Washington)

I found that I could search for people and learned that my old friend Jack who had moved to Missouri in 1997 was back living in Vancouver at his parent's home until he found a house to rent. I called him and we caught up on things that had happened since we last spoke.

He knew René had died having kept in touch with him also, and told me that we needed to get together and talk. I made plans to drive to Vancouver and while there catching up, he said that I was the last one left with so much knowledge about the Sasquatch and all that René had taught me, and that I needed to start writing so that the knowledge would not be lost. I had absolutely no idea how to write. I had written papers in college but a book was an entirely different story. I was a prolific reader, and had to think about how to begin making a book.

(Bear claw marks on a tree Olympic Mountains Washington State)

I thought about the various styles of books I had read and started making a manuscript. It was a very slow process and I did not even have a direction planned. I asked myself, "Just what do I want to tell prospective readers"?

I knew a lot of things about the Sasquatch, and did not want to say the same old things everyone else had written— mainly witness encounter stories. I decided on basically explaining the topic of Bigfoot, and that it was serious and matters as a subject of legitimate inquiry and why I believed this to be the case.

I wanted to demonstrate the historical significance, walking a reader from ancient accounts through current times. I aimed to show continuity and demonstrate that it was not just something concocted in the 1950s, then to briefly discuss evidence. It became something for beginners, but also those who were familiar with the topic. It was not an easy process by any means!

(William Jevning by the Carbon River where he spent much time as a young boy)

Jack was a business owner at this time and doing well financially, and he told me he would financially back the project and pay for editing and associated costs. My part was to create it.
Jack's heart was in the right place, but he kept me on a tight time frame, and never having written anything before I was clueless most of the time and I made frequent trips to Vancouver for whirlwind discussions and planning for the book.

The first editor wanted to entirely change the manuscript to the point I did not even recognize it, and we eventually fired him. I insisted that I was crafting it a particular way and did not want it changed, and the new more professional editor agreed. She even told me I was a natural at writing. So during this process Jack and I also started doing field work together once again, but on a much expanded level.

I was no stranger to working regions from British Columbia to California, but with Jack financially backing my work now we stepped up the efforts. It was during this time we started working an area in Northern California I commonly refer to as area #4.

Area #4 is a place far from where anyone goes to, and I am still working the area so will not divulge its location. I almost immediately started finding plenty of Sasquatch evidence, tracks, snapped and twisted trees and scat, lots of scat.

I put some of this in the book, and just as I had completed the manuscript and we went through the editing process, Jack told me he was retiring and moving to Ecuador.

I cannot complain about his leaving. He finished raising his two sons and had earned his retirement. The book was finished except for the cover and it was left with me after Jack had gone. I had no idea what to do with it! Eventually I was able to publish it, and recently retired the original version, mainly because going between two editors and being out of my hands, often there was some information missing and images I had wanted to improve. So I retired the original, made my changes and the second is an edition I am much more pleased with, which is currently available.

In 2003 I decided to make a new incarnation of my field work and research team, naming it the Jevning Research Group. I adopted a compass as my logo; originally this was a joke because there is no direction in the subject of Bigfoot with people getting involved and being all over the map on this topic. I grew to like the symbol and think it appropriate.

(At camp during a field expedition into the Olympic Mountains of Washington State)

I have done much work since 2003. I will not go into it all, but I have done television and radio since then, and have my own website, williamjevning.com. The Jevning Research Group has grown into a large network of researchers today so I am hopeful about the future and of the subject of the Sasquatch.

To date I have written eleven books with several more in the works. With luck and backing maybe I will yet resolve the question of the existence of the Sasquatch. Time will tell.

The following are some of my photographs during the time I wrote Notes From the Field, Tracking North Americas Sasquatch:

(William Jevning on the Quinault River)

(William Jevning interviewing Al Hodgson at the China Flat museum in Willow creek California)

(William Jevning in the field near Carbonado Washington)

(My photograph of the wood carving made by Jim McClarin at Willow Creek, California)

(Black bear tracks from Mt. Adams)

(William Jevning Hiking and searching the William O. Douglas wilderness Washington State)

(Former members of the PSCIT Board of Directors, Don Turner, Carol Szimonisz, William Jevning and Jack Livingston, at a meeting to help me with my work on the book)

(Another expedition to the Quinault River Olympic Mountains Washington)

(William Jevning in the field at Mt. Adams Washington)

(Entering Area #4 Northern California)

(Twisted and broken tree Area #4 Northern California)

(Our camp near Area #4)

(Camp on an expedition near the Glacier View wilderness near Mt. Rainier Washington)

(Field expedition near Copper Creek, Mt. Rainier Washington)

(William Jevning in the field Estacada Oregon)

(William Jevning not far from where the elk were killed in 1980, this is close to the location of a place we called the alder flats above Wilkeson Washington)

(Rene' Dahinden and William Jevning Vancouver Washington 1992)

(William Jevning next to a friend along the highway leading to Mt. St. Helens visitor center)

(Rene' Dahinden trying to look inconspicuous near the same statue)

(Camp while in the field near Mt. Adams)

(William Jevning and Jack Livingston conducting search near Mt. Adams Washington)

(William Jevning next to the Puyallup River, across the water is the property we called "Old Dick's property"

(William Jevning relaxing at camp in the Olympic Mountains)

(PCSIT brochure)

August 26, 1989: Two teenage girls went out to give a horse grain when a large manlike creature walked up to within ten feet, they ran to the house, the creature followed, they banged pots and pans to scare it away. They said it only looked at them with "a bored look on its face" and walked back to the forest.

September 16, 1989: Four men encountered two large manlike creatures in the woods near the Yacolt Fire Station.

November 20, 1989: Three PCSIT field crew members saw a Sasquatch near the CC Landon Road while conducting routine survey's of the area.

We know based on the many eyewitness accounts we have investigated and the PCSIT staff's own encounters, that many more sasquatch encounters have yet to be revealed.

It is unfortunate that those who usually ridicule others, know the least about which they are scoffing and have the most closed minds.

If this were not the normal situation when people simply tell others about something they have seen, then this issue would have been properly dealt with 40 years ago.

The map shows approximately where Yacolt is located in the State of Washington, and roughly where some of the Sasquatch encounters took place on the blow-up portion.

Again, these may look like a concentration of activity in one area, this represents known encounters since 1925 and when compared to other areas we have researched, it is a common representation.

Anyone having any information from this or other areas who would be willing to speak with an investigator, can write or phone. The information, no matter how little would be very useful. It will be treated with complete confidentiality and assist us in gaining a better idea of what is actually happening in these places.

- *A Word from the President, William Jevning:*

I would first of all like to thank everyone who has responded to our requests for information concerning sasquatch related incidents. The information can only come from you who live, work and spend recreation time in the areas we are researching.

The information we receive vastly helps us put the picture of this issue together much like a jigsaw puzzle. Unlike a puzzle, we do not have a complete picture to go by in assembling this research, this is one reason our task is so difficult, without your help it cannot be done.

With your help by relating any information you may have, no matter how trivial you may think it is, we will resolve this issue.

We have many of the puzzle pieces already, someone may know a small but very key piece and not realize it. Do not be afraid to contact us and talk with us, our guarantee is confidentiality, and you can be certain that no one will laugh at what you

(Page from the PSCIT monthly newsletter titled "Notes From the Field" that was the inspiration for my first book Notes From the Field, Tracking North Americas Sasquatch")

ABOUT THE AUTHOR

William Jevning has been a field investigator and researcher of the creatures known as Sasquatch since 1972. Previous books are Notes From the Field, Tracking North America's Sasquatch, Haunted Valley and In Search of the Unknown and The Minnesota Iceman. He has also been a guest and co-host of many radio and podcast shows and participated in the History channel television show Americas Book of Secrets, the Mystery of Bigfoot.

Made in the USA
Monee, IL
18 August 2022